Everyone's Book of Classic CARS

BC 8449

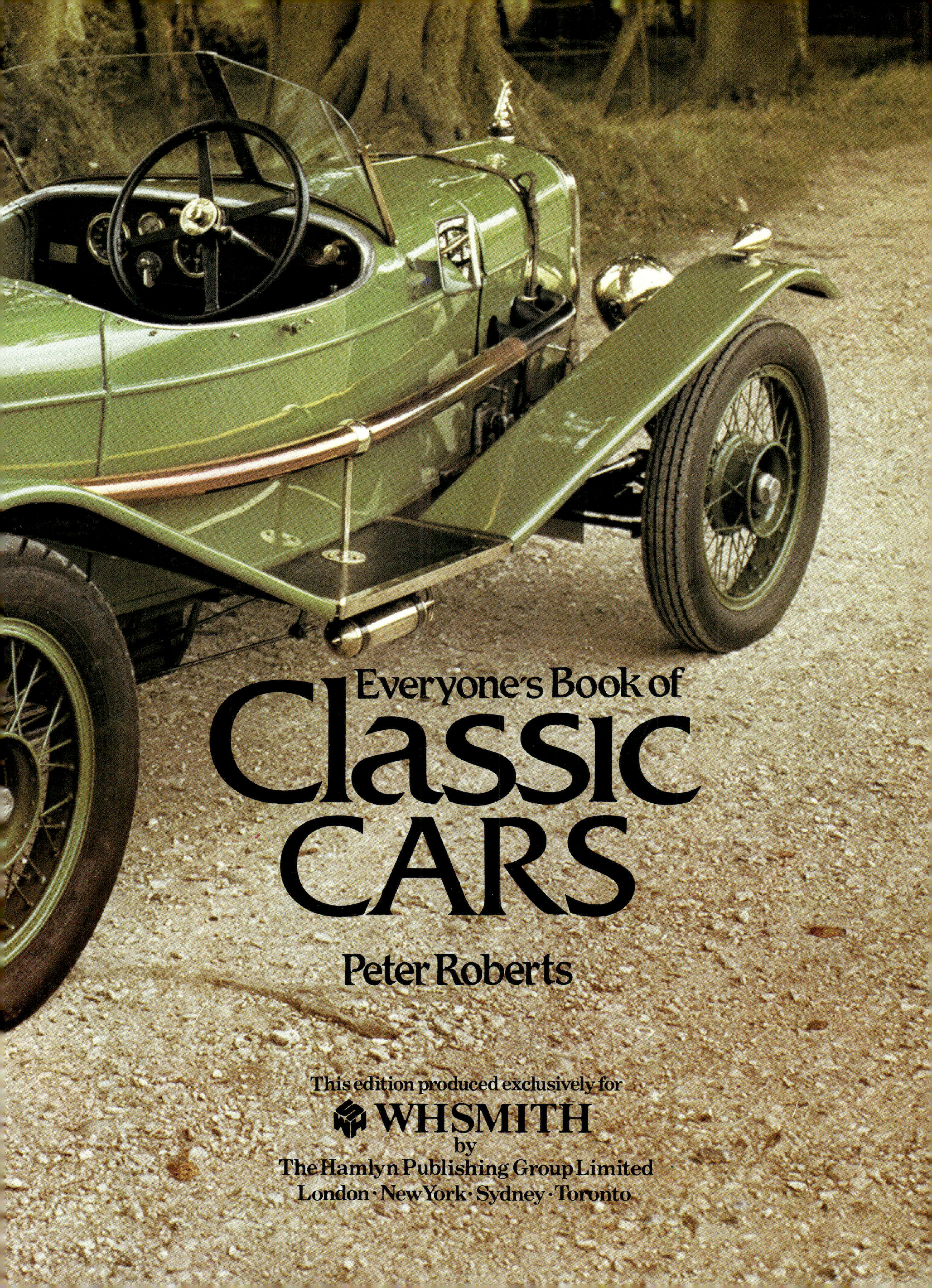

Everyone's Book of

Classic CARS

Peter Roberts

This edition produced exclusively for
WHSMITH
by
The Hamlyn Publishing Group Limited
London · New York · Sydney · Toronto

This edition published exclusively for W H Smith

Published by
The Hamlyn Publishing Group Limited
London · New York · Sydney · Toronto
Astronaut House, Feltham, Middlesex, England

Third impression 1983
ISBN 0 600 37440 8

Phototypeset by Tradespools Limited, Frome, Somerset

Printed in Italy

The publishers and author would like to express their gratitude to the following organizations and persons for their help in research and for supplying photographs for this book, and for valuable editorial assistance.

Adam Opel AG., Russelsheim am Main; Alfa Romeo SpA, Milan; Aston Martin Lagonda (1975) Ltd.; Automobiles Citröen, Paris; Automobiles Peugeot SA, Paris; Mr. Michael Banfield; Mr. Cecil Bendall; British Leyland UK Ltd.; M. Phillipe Charbonneaux; Chrysler Corporation, Detroit; Citröen Cars Ltd.; Daimler-Benz AG., Stuttgart; Fiat Motor Company (UK) Ltd.; Ford Motor Company Ltd. and Mrs. Sheila Knapman; Ford US of Dearborn; General Motors Ltd.; Lips Autotron at Drunen, Holland; Lotus Group of Companies; Maranello Concessionaires Ltd.; Mercedes-Benz (United Kingdom) Ltd.; Mrs Karin Mitra; Musée de l'Automobile Française; Panther Westwinds Ltd.; Porsche Cars Great Britain Ltd.; Régie Nationale des Usines Renault, France; Renault Ltd.; Rolls-Royce Motors Ltd.; The Shuttleworth Collection; Vauxhall Motors Ltd.

Thanks also to Mr. Michael Worthington-Williams of Sotheby's for his able help in matters editorial and research.

Photographic acknowledgments
Neill Bruce: pages 43 top, 48 top, 51 top, 60, 66 top, 71 bottom right, 76. Nicky Wright: pages 53, 55, 56 right, 59 bottom, 64 top, 64 bottom, 72 top. (Photographs on front cover and pages 55, 56 right and 59 bottom are also by permission of the Auburn-Cord-Duesenberg Museum).
All other photographs are from the Peter Roberts Collection.

Front cover: 1931 Duesenberg J (Nicky Wright)
Back cover: Porsche Turbo (Maurice Rowe)
Endpapers: Lamborghini Countach (Maurice Rowe)
Title spread: 1923 Alvis Ducksback (Neill Bruce)
Contents spread: Mercedes-Benz 7·1-litre SSK (Peter Roberts Collection)

Contents

The Dawn of the Automobile Age (1885-1904)

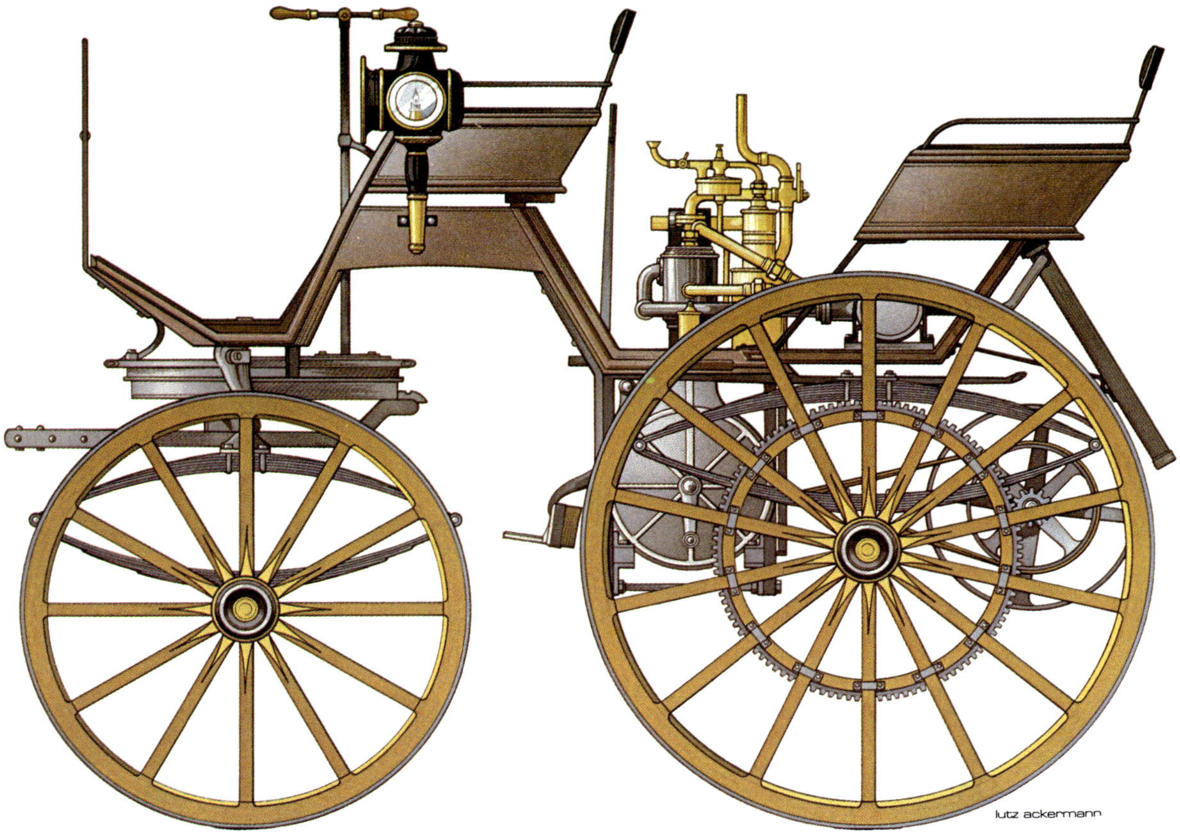

This light carriage, fitted with a single cylinder engine in 1886, was Daimler's first true automobile. Later he installed his engine in a river launch and an airship.

The scene is a small town in Germany, the time New Year's Eve, a century ago, in the small dining-room of a modest suburban house. Outside in the yard is a small workshop. The year 1880 is just ten minutes away.

The pretty young housewife turns to her 35-year-old engineer husband: 'Karl, let's go over to the workshop and try our luck once more. Something tells me that we should try one last time.'

Bertha and Karl Benz walk over to the little engine he had been building over the past weeks. Benz seizes the cranking handle, turns it, his heart pounding. . .

This is what Karl Benz, father of the first practicable automobile, said years later of the next historic moments. 'The engine started to go put-put-put and the music of the future sounded with regular rhythm. We both listened to it for a full hour, never tiring of the single note of its song. The two-cycle engine performed as no magic flute in the world ever had. . .for we knew the greatest kind of happiness that evening in our poor workshop. Suddenly the bells began to ring out, ringing in an era which was to take on a new heartbeat from the use of this engine.'

Six years later Benz's first automobile had been built and tested in his backyard shop, using an improved version of the crude power unit that had been born on that New Year's Eve.

The first Benz 'Patent Motor Car' was considerably indebted to the cycle industry that was flourishing during that period. Cycle-type tubing

Left: Daimler's first motor vehicle was this single-track machine, completed in 1885. This was the inspiration for his four-wheeled motored carriage of the following year.

was used for the frame, cycle-type spoked wheels were used and chains and linkages of cycle design were employed in the Benz lightweight three-wheeler. The tiny ⅔ hp engine was a miracle of invention and innovation; its single cylinder had a slide inlet valve something like today's two-stroke unit, cooling was by water evaporation in a jacket round the cylinder. Karl Benz developed an ignition system which used coil, sparkplug and battery. The drive to the road wheels was by pulley and belt to a transmission shaft which had a differential, and by moving the belt from idler to working pulley the drive could be engaged and disengaged. Engine speed was controlled by changing the quantity of air admitted to the mixture of air/petrol vapour from a surface-type carburettor. Simple and rough were these systems, but how curiously modern those technical terms still sound to us a hundred years later.

Meanwhile, down at Cannstatt, a few miles away from the Benz home in Mannheim, another engineer was busy in his workshop. Gottlieb Daimler, 51 years old, had been working on a petrol engine for some time, and had made a compact, water-cooled, four-stroke, single-cylinder unit which he patented on 3 April 1885. Daimler's objective was to design an engine that could power any type of vehicle (unlike the Benz engine, which was designed specifically to propel a road carriage), and his first choice of experimental test-bed transport was a hobby-horse type motorcycle. He fitted it with his ½ hp motor, and trundled around his garden in the unwieldy 'single-track' machine. His loyal assistant and lifelong friend Wilhelm Maybach is credited with building this unit, one small enough to fit into the cramped space between the wheels of the cycle.

Daimler's second attempt at self-propelled road transport was more ambitious. He ordered an open family coach to be delivered to his premises, keeping his real intentions secret by saying that he wanted to give his wife a surprise birthday gift. Clandestinely he fitted a 1·1-hp engine into the rear floor-space of the carriage.

This unit was fired by hot tube ignition, a simpler method than that used by Benz but one which was less positive in action. The engine was mounted on rubber shock absorbers, the water fan-cooled (at first) and the gas-air mix preheated by exhaust gas – all pointers to the future and all on the world's joint-first automobile.

Above: The first machine built by German engineer Karl Benz rattled out of the workshop on 3 July 1886. Benz's 'Patent Motor Car' owed much to cycle design as can be seen from its lightweight construction.

Daimler, visionary and pragmatist, fitted his engine into a river launch, running it up and down the local waterway, the Neckar, to everyone's astonishment. Then in 1887 he installed one in a streetcar, and later fitted out an airship which had been experimenting with steam engines, with a propeller-driving gas unit which took the dirigible on a successful first flight of 4 km (2½ miles).

Thus the first internal combustion engine-powered air and marine craft were launched, and the first shaky automobiles rattled over the cobbled roads of Germany – the ancestors, for good or ill, of today's teeming millions of vehicles.

Progress was slow at first, but by 1888 Benz had published his first advertisement, and the first sales had taken place. When Benz overcame the tricky technical problem of steering by the two front wheels (he had designed his first car as a three-wheeler to avoid the problems associated with two-wheel steering), had made his first four-wheeler, the Viktoria, and had moved on to the Velo, the world's first small standard production car, he moved into an early form of mass production. In 1894, the first year of the Velo, Benz sold 67 machines. The following year he sold 135; by 1900 annual sales had risen to 603. Then the result of a dramatic event in the automotive world hit the Benz sales graph.

In 1900 Daimler, who by this time was selling motor cars in several European countries as well as in the United States of America, had been engaged in motor sport for several seasons. That year his French agent Emil Jellinek, former racing driver and diplomat, had found the 23-hp Daimler from Canstatt far too unwieldy for hill-climbing competition. A driver had been killed conducting one of his cars up the La Turbie hill-climb contest near Monte Carlo. Jellinek persuaded the Daimler company (Gottlieb himself had died in 1900) to make a lower, longer model of increased horsepower.

Thus a new car entered the scene, and a new motor age opened. The 1901 Daimler product had so many advanced features that it rendered every other motor car obsolete overnight. With a beefy 35 hp from its 5·9-litre, four-cylinder engine fed by jet carburettors, the car scooped the field during the Spring meeting at Nice in March 1901, winning almost every prize offered in competition.

Emil Jellinek had promised to sell a first series of 36 of the new cars if they were made to his specifications. His order book was soon crammed with the addresses of would-be buyers.

One factor, apart from the obvious technical advances of the car, aided its universal sales success. There had been a certain amount of German sabre-rattling in Europe at the time, and Jellinek had insisted that the car be given a less Germanic name than Daimler. It was agreed that the 1901 model should be named after his 11-year-old daughter Mercédès. As French writer Paul Mayan, secretary general of the French Automobile Club was forced to admit in print after the debut of the new car at Nice, 'We are now in the era of the Mercedes', which must have been galling for a patriot whose allegiances would have been fiercely with the French Mors or the great Panhard et Levassor motor company which had hitherto been leading the automobile world in reputation.

Above: The Mercedes, a car named after the daughter of Daimler's French agent, appeared in 1901 and rendered all other motor vehicles obsolete overnight. This is the 1902 Mercedes-Simplex tourer, its direct successor.

Above right: The Stahlradwagen or 'wire-wheeler' car of 1889 was fitted with a twin-cylinder engine, and was Daimler's first vehicle to break away from carriage-building tradition.

Right: The bizarre Schlumpf collection of antique cars in Mulhouse, France, was discovered after many years of being kept secret by its owners. This is a nineteenth century steam vehicle that as yet has defied identification.

FRANCE TAKES THE LEAD

The experimental age of pre-motoring development had been mainly in Germany. Marcus had made an early self-propelled cart driven by an internal combustion engine, Nikolaus Otto produced a four-stroke gas engine as early as 1876 in his Deutz factory (where the technical director has been the young Gottlieb Daimler) and the brilliant work of Benz and Daimler in the last years of the nineteenth century pointed to the continuance of German leadership in the field of road transport. However, the heartland of the motor industry had moved to France from Germany, the country of its nativity, during the 1890s when licences to make the engine were granted by Daimler to a number of French companies.

Perhaps the French had more commercial enterprise for they rapidly stole the scene from their neighbouring country during the years of automobile development at the fold of the centuries, forming automobile

associations, mounting sporting contests, creating speed records, forging new companies, marketing new vehicles. So enthusiastic were both manufacturers and buyers in France during those formative motoring years that the 'official' language of motoring and motor sport is still French.

Certainly the 1889 Paris Universal Exposition had created a French interest in mechanical power. Daimler was there with his new 'Steel Wheeler' and some of his engines. The spindly car, built on cycle lines, was not of great interest to the crowds packing the exhibition, but the uses to which his engine could be put caught the imagination of several people, including a lady and her constant companion, a Monsieur Emile Levassor. The lady's husband had recently died and his last advice had been that she foster their current business relationship with Herr Daimler – he had been a representative of Daimler's company – and that she take over his recently agreed rights to make the latest Daimler engine under licence in France.

Mme. Sarazin's companion Levassor married her, and by 1890, with his partner René Panhard, had made and sold the first of the engines. Panhard and Levassor had given permission to Armand Peugeot to install the engine in a carriage, although in fact they produced their first automobile slightly earlier than Peugeot's 1891 quadricycle.

History has more to say about the first Panhard than of the Peugeot. The original Panhard had its engine midships, under the seats. But for several reasons the unit was moved to the front of the car (the official explanation was that the steering system would be more stable; an unofficial one was that the ladies found the seats too hot) thus unconsciously setting down the conventional layout of today's motor cars – front engine, followed by clutch, gearbox and drive to the road wheels. Its gearbox was exposed to the elements and gave a choice of indirect ratios, bevel drive to a cross-shaft and side chains, a design that became known as the *Systeme Panhard*. It was based on Levassor's experience of lathe-belt speed-change systems, and as he said: '*C'est brusque et brutal, mais ça marche*!' – 'It's rough and brutal, but it works!'

The Peugeot company had long been in the business of making metal parts for anything from crinolines to bicycles. Armand, scion of the ancient family, had been studying engineering in England and had foreseen the future of motor transport. His first motor carriage employed the Daimler V-twin power unit at the rear with its water cooled through the tubular, cycle-type frame on which the car was built. The steering was by cycle-type handlebar.

A Peugeot 4-place *vis-a-vis* (passengers and driver sit face-to-face) made the first-ever long-distance, cross-country run in 1891 when it was driven 2,047 km (1,272 miles) following the Paris-Brest-Paris cycle race 'without a moment's trouble' at a speed of around 15 km/h (9 mph). Peugeot, with their cycle-designing experience, seemed to be able to produce more reliable vehicles than the heavier, more cumbersome Panhards of the late Nineties, and the cars had markedly better driving characteristics, partly due to their superior suspension.

During the early twentieth century large numbers of small manufacturers mushroomed (the total number of U.S. manufacturers from

A Panhard-Levassor of about 1895, with twin-cylinder engine. By this time the *Systeme Panhard* was being copied by many other manufacturers.

De Dion-Bouton, circa 1899; a famous advertisement showing a Model D voiturette, two tricycles, a quadricycle and a De Dion steam bus.

pioneering days to the present stands at well over 2,000). Many were merely backyard tinkerers who soon sank without trace, but some, with perhaps a past history of engineering in cycles, sewing machines or other light machinery, began to flourish, including those whose background had been electrical and who were pushing ahead with battery-powered transport, or those in the locomotive field who favoured steam as the motive power for road vehicles.

De Dion-Bouton was one of the most popular French makes of the time. Born of an alliance between nobleman Comte de Dion and steam engineer Georges Bouton, the company first made steam vehicles – *automobiles à vapeur* – but changed to petrol in 1883. De Dion tricycles and voiturettes (light cars) using a high-speed engine of unusual efficiency were more numerous than any other marque at the turn of the century, and European city streets rang to the warning bells of the little French 'Ding-Dongs'.

France was making a bewildering variety of automobiles by the time Edward, Prince of Wales reached the English throne, from the little Decauville to the massive Mors and De Dietrich racers. Darracq, Delaunay-Belleville, Gobron-Brillie were representatives of the more aristocratic products, while dozens of cheaper cars were made in high hopes of mass sales.

Henry Ford had made his first internal combustion engine back in 1893. His first car, powered by the unit seen here, could reach a top speed of 25 mph.

The name of Louis Renault appears early in this field. At 21 this young son of a middle-class French family living at Billancourt in suburban Paris was annoying the neighbours by working at all hours in the shed at the bottom of the garden. Firstly adapting a De Dion tricycle, then building a four-wheeler of his own in 1899, he persuaded a few friends to order his new product. Within a year the three brothers, Louis, Marcel and Fernand, had formed a company. Six months later they had sold 60 vehicles and had made the world's first saloon car. Louis Renault was acknowledged France's first captain of industry by the time he was 29 and was made a Chevalier de la Legion d'Honneur in 1906. His company was turning out over 10,000 cars a year by 1914 and had opened branches all over the world.

AMERICAN PROGRESS

In the U.S.A. the automotive industry at the turn of the century still had its feet firmly stuck in the muddy waters of fundamental automotive problems, such as making a choice between steam, electric or gasoline power, and in the equally sticky mud of its appalling road system. The general opinion of American roads of that time led most international travellers to opine that it was easier to go around the North American landmass than over it.

However, although U.S. pioneers lagged far behind the technical standards of French and German manufacturers, suffering from lack of knowledge of what was being developed in Europe, they plunged into this new industry with typical American application and energy.

The brothers Duryea had built a Benz-based machine, successfully tested it, and had even raced it by 1895. Electric vehicles had been built and sold, but their range was restricted to city limits by their need for regular battery charging. Steam vehicles had been tried a generation earlier, and still had a large following. Hiram Maxim had offered his internal-combustion-engined lightweight car to a Colonel Albert Pope, who turned it down as 'people wouldn't want to sit on top of an explosion'. Colonel Pope started making electric cars with considerable success, a success that ironically led him eventually into the gas-engined field – to build a veritable empire of early vehicles.

But like the British motor industry, which was held back in pioneer days by diehard opinion and legislation, American automobile manufacturing was slowed down by an historic legal hiccup. In 1895 lawyer George Selden had cornered the patent on nearly every part of the early automobile, and for 16 years, until Henry Ford upset his patents claim in a lawsuit that became part of history, Selden played cat among the pigeons in the infant American industry, demanding that new auto makers pay up or cease business.

Farmer's boy Henry Ford had been in business for some time by 1911, the time of the great lawsuit, and had built his first engine back in 1893 out of gas-piping and wire. His first car, a lightweight quadricycle owing much to bicycle and soapbox ancestry, appeared three years later and was remarkable in that its two tiny cylinders could take it along at up to 25 mph although it had several small disadvantages such as no brakes or reverse gear!

By the turn of the century, every American with the remotest claim to manufacture had plunged, it seemed, into the industry.

The Stanley brothers, who produced the classic steam car, were formerly photographic plate makers; George Pierce, later to produce the acclaimed Pierce-Arrow, had made birdcages and bicycles, young Ransom Olds was a bookkeeper at his father's repair works, James Packard had manufactured electric cables, and David Buick was a plumber. Many of the early twentieth-century auto makers of America, quality companies and monkeywrench engineers alike, went to the wall before many years had passed (and many more vanished during the Depression of 1929), but some went on to found marques that would become household names of the period – makes like Packard, Marmon, Cadillac, Buick, Stutz, Locomobile, and so many others.

Many made light, high-wheeled autos, ranch or farm buggies that became the classic form of American car in a period when the roads were

Formal dress in a De Dion-Bouton! This one is an 1899 voiturette with a water-cooled unit mounted at the rear, and two-speed transmission.

The first Renault voiturette, with a De Dion-Bouton air-cooled engine mounted at the front. This car had shaft drive and 3-speed transmission.

still deep in mire. One such was the Holsman, a two-cylinder lightweight with an ability to negotiate rough tracks and to ford shallow rivers. This model sold in considerable numbers in the Mid-west where it suited road conditions admirably.

But the most popular American car in the buggy tradition was the little Curved Dash Olds. Produced by Ransom Eli Olds in Lancing, Michigan, the 'Merry Olds' was a classic in many ways at the time of its introduction, and has certainly become a sought-after classic today.

The Curved Dash brought motoring to many who would not have been able to afford a more luxurious vehicle. Simplicity was the keynote of the car, and Olds brought the first rudiments of quantity production to its manufacture. He also introduced mass-marketing techniques for a low-price car, giving confidence to investors who had earlier considered the auto industry a somewhat hare-brained gamble. The Olds works also proved to be a training ground for many budding engineers who were to move on to their own auto-making businesses. The Curved Dash, with its tiller steering and single-cylinder unit giving 'one chug per telegraph pole' and 7 hp, rolled into the history books by rapidly topping the sales charts (425 in 1901, 5508 in 1904) and by performing a number of marathon coast-to-coast trips watched by an admiring American public.

A car that came to be called the 'Standard of the World' was the modestly-launched Cadillac. In 1903 Henry Leland, who had been associated with Ford's early ventures, took over the car-making side of the company that Henry had left, renaming the product Cadillac. It had a similar overall design to Ford's Model A, the first car to be produced by his new Ford Motor Company, but under Leland's meticulous control soon became known for its superb engineering quality. At a time when every automobile was built individually with parts machined and trimmed to match only a single-partner unit, Leland's insistence on the accuracy of each part of the vehicle in manufacturing allowed an astonishing demonstration of precision when three similar Cadillacs were completely dismantled, the parts scrambled and reassembled, and the cars immediately driven.

A turn in the Bois for madame in her Peugeot Victoria, circa 1892. It was powered by a 1282-cc engine built under Daimler licence.

Ford No 1, completed in June 1896 and powered by a twin-cylinder, chain-driven engine which was surprisingly reliable.

BRITAIN ON THE ROAD

Britain, until 1896 still under the restrictive laws which stated that motor vehicles should not exceed 4 mph, was also, like America, lagging behind the European motor world. French and German enthusiasts were enjoying the new sport of motor racing whilst in Britain it was still necessary for a footman to precede the vehicle. Numbers of influential 'autocarists' were lobbying in high places, however. The Prince of Wales had been out for a spin in a new British Daimler and had caught motoring fever. Others in Parliament were persuaded to the cause. But most authority was still firmly in the horse era, and both the police and the courts were heavily biased against the motorist. One sometimes cannot help agreeing with some of the diehards though, for early automobiles and their drivers were often very real dangers to children and animals, and were certainly polluters of the countryside and urban regions with their great plumes of following dust.

The country had its share of forward-looking engineers, however, men who had seen developments in Germany and in France and were moving

Durkopp of Germany made this little Panhard-based two-cylinder 8-hp vehicle in 1901. They sold as Watsonias in Britain, and under the name of Canello-Durkopp in France.

towards motor manufacture as fast as they could. F. W. Lanchester, a Midlands engineer, made his first prototype in 1895, a car that, like that of Karl Benz, he had built from the ground up without reference to traditional carriage construction. The result was a vehicle with a centrally mounted, horizontally-opposed, twin-cylinder, air-cooled engine which had an unusually smooth and quiet operation. His early cars, and later Lanchesters, were always unorthodox and advanced for their time, and whilst the company finally failed, it made numbers of important pioneering steps in automotive engineering.

Several small companies had begun to copy Continental models before 1900, but one of the earliest serious entrants into the field was Herbert Austin, a young man engaged in the somewhat seasonal work of managing the Wolseley Sheep Shearing Machine Company. To fill in he began making two light three-wheeler vehicles based on the Léon Bollée he had seen on a business trip to France.

Austin's work for Wolseley resulted in the first four-wheeled model being offered to the British public in 1899, and Royalty, in the person of Queen Alexandra, purchased a Wolseley in 1903.

Austin himself however had a disagreement with his directors about design – he wanted to produce engines with horizontal cylinders while his company wished to build vertical engines – which led him to set up his own company in 1906, the Austin Motor Company at Longbridge near Birmingham in the industrial Midlands of England.

The Daimler Motor Syndicate had been formed as far back as 1893 at Coventry under Gottlieb Daimler's patent. It produced its first cars in 1896 under the banner of entrepreneur H. J. Lawson who, like Selden in America, had attempted (eventually unsuccessfully) to corner the automotive market in Great Britain.

The turn of the century saw the first glimmer of an emerging industry, rather than an unrelated number of engineering projects. Napier, well known as printing machinery makers, had brought out their first series model, based on Panhard-Levassor design; American Wilbur Gun from Springfield, Ohio had started building small tri-cars in England, reliable little vehicles that evolved into the Lagonda sports cars that captured a large *aficionado* market in later years. Another marque born then was Sunbeam, product of the firm of John Marsden, the tinplate and japanning company.

The marine engineering company, Vauxhall Ironworks Ltd., on the banks of the Thames in west London, was also casting about for other

Below: First Fiat. This poster of Fiat's initial product was seen in 1899. The 3½ hp model was soon succeeded by front-engined models.

Bottom: Mass production pioneer. The Curved Dash Oldsmobile was the first quantity-built automobile in the U.S.A. This is a 1901 model.

work, and entered the automotive field, one of the few pioneering companies in Britain still producing today a car bearing its original name. The same year as the Vauxhall made its debut another small company, which had been in the steam truck business for a time, decided that the fashion in transport was changing rapidly and produced the first Thornycroft internal combustion engined car, later models of which showed distinct similarities to the early Mercedes as did a high proportion of larger cars made during the first five years of the twentieth century.

Engineer Henry Royce had driven his first motor car out of his workyard in Manchester in April 1904 and had offered the Press a spin. All who where privileged to ride in the little 1·8-litre two-cylinder, shaft-driven Royce eulogized about its silence and smooth operation, and as one scribe wrote: 'The Royce car made all others sound like an avalanche of tea-trays!' When later in 1904 Royce met the Hon. Charles Rolls who had been selling French cars in London, the scene was set for the birth of the 'Best Car in the World', as the world called the Rolls-Royce.

By 1904–5, the end of the truly pioneering era, much of Europe was deeply involved in the motor industry. Germany had established a high reputation with Benz and Daimler, still separate companies, making advanced four-cylinder cars (the first six-cylinder Daimler, the Mercedes tourer, appeared in 1906). Opel, formed as a sewing machine and cycle company, had broken into the field, first with an alliance with the French Darracq company then in 1902 making their own complete cars in the family works in Rüsselsheim near Frankfurt am Main. Opel's first home product, a tidy little 10-hp with a two-cylinder 1884-cc power unit, was to set them on the road that took them into the lead as German motor manufacturers before the disastrous Depression of the early 1930s.

ITALY'S EMPIRE

South of the Alps, Italy, largely unsung as one of the real pioneers of road transport, started relatively late in the motor field. One of the oldest and most respected marques, Isotta-Fraschini, dates from 1899 (and continued for exactly half a century, during which time the car acquired an almost legendary reputation for sportive luxury), and other Italian classics – Lancia, Alfa Romeo, Maserati, Ferrari, Lamborghini – appeared on the scene considerably later.

The Italian scene is, and always has been, dominated by the giant Fiat concern in Turin. Fiat – *Fabbrica Italiana di Automobili Torino* – was one

Ideally suited to early U.S. Middle West roads, this high-wheeled Holsman Rambler of 1902 was made in Chicago. Its simple twin-cylinder unit and spartan construction enabled it to be sold cheaply in the U.S.A.

Above: 'The Standard of the World' was the claim of the Cadillac Motor Car Co. of Detroit. This 1906 10 hp Model K two-seater could be transformed into a four-place tourer.

Left: Lanchester 1903, 12 hp and extremely unorthodox with its centrally-mounted engine and side-tiller steering.

man's dream, backed by a syndicate of wealthy Turin financiers. Giovanni Agnelli had been a junior officer in the Italian cavalry, had left the army, persuaded some local bankers to buy an old cycle works and there made his first automobile in 1899.

Agnelli's small car, a mite old-fashioned for the day with its flat-twin rear engine, face-to-face seating and huge Victoria canopy, sold 20 examples, a healthy start for the infant company. Fiat fortunes increased with the speed of a runaway horse. By 1903 the company was producing trucks as well as several car models. Two years later Fiat was in ship-building and ball-bearings, and was the country's leading manufacturer of marine engines. Aircraft and locomotives completed the company's coverage of the transport world. Today Fiat is one of the largest and most powerful manufacturers *of any type* in the world.

Always well advanced in engineering and manufacturing technology, the motor makers of Italy have continuously produced cars that have qualified for the accolade of classic, confirming the old saw that 'Every Italian is an engineer and every engineer an artist'.

Opel-Darracq 1902. This year the German company imported Darracq chassis, mounting their own bodies on the cars. Single cylinder; top speed 45 km/h.

Three years into the new century saw dramatic developments in the motor industry – and in the motoring habit. The years of experimentation were not over, yet much of car design had become established. By 1903–4 most well-constructed cars were using the *Systeme Panhard* with the engine settled down in front of the car and the drive train through clutch and box and chains or shaft to a differential and rear road wheels.

Motoring was still, in Europe at least, a hobby for the rich, something 'to indulge in between fishing and ping-pong' as one contemporary journal put it, and a drive into the country was an accepted weekend

pastime – as long as it didn't interfere with hunting.

Only in America was the new transportation taken seriously by the public, and here the buyers were, more often than not, doctors or travelling salesmen whose gratitude to the motor industry was unbounded after years in the saddle or on a bouncing buckboard.

Most cars were still open to the elements. Designers were still so occupied with getting the car to its destination that weather protection was virtually ignored, except for the type of clothes recommended for travelling. But although the man-in-the-street (even in the streets of London or New York) would be quite likely to traverse town without seeing a single car, automobile development was moving forward rapidly and irreversibly.

In 1903 world production of cars was 61,927, a fantastic number for an industry only a bare ten years old, and considering that the horse was still king of the road in the minds of a vast majority of the world's population. France still led the field commercially, producing some 30,000 cars that year, a figure that indicated that she was expanding eagerly into the new age. Technically, the 1901 Mercedes had given manufacturers a jolt, one that soon became a shot-in-the-arm as engineers buckled down to the task of copying and improving on the advanced German machinery.

Complete cars were not yet made and sold as single items. The chassis was purchased, then the form of body discussed with the coachmakers, built carefully and laboriously by hand, and fitted on to the frame. Naturally most of the bodybuilders came directly from the horse-drawn coachbuilding trade, with the result that car shapes, although long past the horseless carriage stage, were often still adorned with the weighty curlicues and decorations of an earlier era. Only the smaller, cheaper vehicles seemed to escape the rather stultifying tradition; with their construction designed for economy and lightness there was little room for décor or Victorian comfort.

Some five years into the new century the aim of motor engineers had changed subtly. One of Europe's foremost writers on the subject has put it more clearly than most. Said Laurence Pomeroy Jnr: 'From 1885 to 1895 men struggled to make the car go. From 1896 to 1905 they contrived to make it go properly. Between 1907 and 1915 they succeeded in making it go beautifully.'

Former steam-lorry makers Thornycroft produced their first car in 1903. This is a late model of that year, with finned tube radiator.

Vauxhall Ironworks Ltd. moved from river craft to cars in 1903. This is their first offering to the public – a 6 hp runabout.

Sporting Challenge

Competition is natural to man. Rapid travel has always been one of his fondest ambitions. The search for personal transport has been another. And when in the course of history science offered the chance for him to compete in a new form of mechanical transport, his enthusiasm knew few bounds.

The first competitive motoring events were designed as endurance tests. Autocars, as many people called them, were wildly unreliable during the formative years of the 1890s, and most early vehicles were not used as serious transport to a destination but were leisure objects, used at weekends by the wealthy for their amusement.

However, by 1894 one man at least, the *chef des informations* of a French magazine *Le Petit Journal* decided to encourage designers and owners by mounting the world's first motor sporting contest, a reliability trial.

From Paris to Rouen was a distance of some 120 rural kilometres (75 miles) and on the morning of 22 July some 21 vehicles, from cumbersome iron-clad steam trucks to light-weight phaetons on their cycle-type chassis and wheels, set off from the Porte Maillot in Paris for the waiting crowds at the ancient cathedral city of Rouen.

Although the winner would be the vehicle that, as the regulations stated, was 'without danger, easily handled by its driver, and of low running costs', the spirit of speed competition was evident before the first batch had left the cobbles of the Porte Maillot. Finally a Panhard and a Peugeot, both using the Daimler V-twin engine, were awarded joint first prizes. The long, nerve-shredding, heart-pounding story of motor racing had begun, with its colour and its victories and tragedies.

Nothing could stop that coterie of sportive Frenchmen from mounting another event the following year...and a dozen more races in the

Peugeot 1895. This model was similar to the one that took part in motor sport's first event, from Paris to Rouen in 1894.

next. . .until national borders had been crossed in the name of motor sport, and Germany and Italy and America had joined in, and finally Britain had found a vehicle worthy of competing with the world's best – the Mors, Mercedes, Panhard, Darracq and De Dietrich cars.

By 1902 capital-to-capital races were the vogue, long-distance marathons that would include dizzy mountain ranges and fast straight sections where speeds were already topping 140 km/h (90 mph).

The great race that year was from Paris to Vienna, with another race running concurrently over the same route, from Paris to Innsbruck in

De Dietrich entered this chain-driven two-seater car in the Paris Madrid race of 1903. At this time the French car was beginning to make its mark in motorsport.

Austria. The longer race of 1,120 km (695 miles) was won by a small Renault car competing in the voiturette class, driven by Marcel Renault against vastly more powerful opposition that included the much-vaunted Mercedes – the car everyone thought should win easily. The Paris-to-Innsbruck race for the Gordon Bennett Trophy, a cup donated by the American newspaper editor of that name and run at the same time, was won by Englishman Selwyn F. Edge in a lightweight 6·5-litre Napier, the first time Napier or any other British car had won an international sporting event.

By this time publicity created by a win in an important race brought customers to the maker of the victorious car, a fact that did not escape the notice of manufacturers who were now putting great energy into producing sporting vehicles – not necessarily for sale to the public – and training racing drivers to handle them.

Racing on public roads soon became too dangerous to be allowed to continue following a number of accidents, particularly in the Paris-Madrid race of 1903. By 1906 speed competitions had been confined to closed roads which could form a continuous circuit.

The first major event to be held under these conditions was at Le Mans in the Sarthe region of France, a race which was the first of today's annual Grand Prix series.

The 103-kilometre (64-mile) circuit was over closed public roads, joined into a rough triangle by a temporary connecting road. The weather was intolerably hot. Earlier, dust had plumed up over the circuit at the slightest breeze. To lay this hazard the *piste* was tarred, later causing an even worse hazard as melting tar sprayed up into the faces of the drivers.

There was a large entry. Thirty-seven teams had brought their cars from all the European manufacturing countries except Britain, who could

This Renault was the surprise winner of the Paris-Vienna race of 1902, with Marcel Renault at the wheel. The 3·7-litre car averaged 62 km/h (39 mph) over the 615-mile race.

not find a suitable vehicle. France put 28 cars on the start-line, amongst which was the Renault team.

The No. 1 driver for the Renault *équipe* was Hungarian Ferenc Szisz; his car a 13-litre (not by any means large for the day when compared with the 18·3-litre Panhards in the race) with shaft-drive, a feature of Renault cars from the earliest days, and one that was not then the accepted method of transmitting power to the road wheels, which were usually driven by heavy, rattling chains. The car also had another innovation – detachable wheel rims. With an expected half-a-dozen punctures during the long race, this quick-change system proved a distinct advantage when most other competitors had to cut off the old tyre, struggle with safety bolts to put on a new one, pump it up and restart the engine. The new rims reduced changing time by vital minutes.

Held under a brassy June sun, the 1906 Grand Prix de l'Automobile Club de France, as the French Grand Prix is still officially called, was a classic that went straight into the record books. Szisz won the long race at

Hispano-Suiza Alfonso 1912. Based on a race-winning voiturette of the same marque, this classically-engineered sports car was best known with a 3·6-litre engine.

Vauxhall's classic model, the Prince Henry of 1914, housed a 3969-cc four-cylinder unit of 25 hp.

a daunting average speed of 101 km/h (62·8 mph), pursued hotly by Nazzaro's 100-hp Fiat, after a two-day drive of almost 13 hours.

The following year the French Grand Prix was run at Dieppe in Normandy, when fortunes were reversed. This was Fiat's great racing year. The first two cars past the flag at the French race in 1907 were of the same marques as the previous year but this time Nazzaro's Fiat led the Renault driven by the Hungarian Szisz. Fiat cars in this, their *annus mirabilis*, also won the other two major races that year, the event over the treacherous roads of Sicily organized by wealthy promoter Vincenzo Florio earlier in the year, and the Kaiserpreis, the race held for touring cars over the Taunus Mountains near Frankfurt in Germany. The little Italian driver Nazzaro, son of a coal merchant, won all three for Fiat, a treble victory that Fiat never surpassed. Twenty-six-year-old Nazzaro's win in the French event was in Fiat's most powerful car, the 16·3-litre, 130-hp, four-cylinder Corso designed by Carlo Cavalli, one-time lawyer turned technical director.

END OF THE MONSTERS

The demise of the multi-litre monsters of the first years of the century came shortly after it was seen that lighter cars with smaller, more efficient engines, could outpace the giants. Even Panhard, one of the founders of the dinosaur breed, began to bring their engines down a few litres, as designers realized that brute power and piston size availed little against efficient design. By 1914, the end of the Edwardian racing era, engines had become far smaller, the winner of the French Grand Prix of that year being a racing Mercedes of just 4·5 litres.

Meanwhile, on the public roads, cars had taken several directions in design. Luxury cars, plush with traditional coachwork leavened by marine influence – portholes, brass speaking-tubes, port and starboard signals were *de rigueur* – vied with spartan tourers and sporting vehicles.

Top right: Descendant of the Prince Henry, the 4·5-litre (later 4·2) Vauxhall 30/98 first made in 1913, was highly successful in competition during the Twenties, yet could safely take the family out for a spin.

Prince Heinrich of Prussia had offered a trophy in 1907 for a trial of cars that could be used on the road, and that could not in any way be termed racing cars. From the third of these events in 1910 emerged two vehicles that were perhaps the first true sports cars. One, an Austro-Daimler designed by Dr. Ferdinand Porsche, housed a 5·7-litre unit, old-fashioned but tough, and the other, a car that was closer to the standard touring car of the day, a 3-litre Vauxhall. The Vauxhall's reliability at the 1910 German trials had caught the eye of the public, and its designer, Laurence Pomeroy, had given it a 75 mph capability. Amateur racers at Britain's only motor racing circuit at Brooklands began to use this model

Above: Few cars are remembered with more affection than the Bentleys of the Twenties. This is a 1929 4398 cc (blown) Bentley at a vintage meeting in Britain.

Fiat 'Mephistopheles' the huge 18·155 cc racer that challenged – and beat – a Napier at Brooklands in 1908. It was rebuilt in 1922 and broke the world speed record in 1924.

(now dubbed the Prince Henry Vauxhall) in stripped-down form, and several single-seaters were made. The Prince Henry created numbers of speed records and captured major rally trophies but could still be marketed as a car in which to take the family out for a Saturday spin.

Vauxhall's Prince Henry was succeeded by the 30/98 (nobody knows to this day why the car was so called) in 1913, but only half a dozen were made before production was halted by the First World War. However, within its single pre-war season the Vauxhall 30/98 had notched up an impressive score of victories in hill-climbing, racing and reliability trials – and this car too could be flexible and accommodating enough to be used with safety for ordinary motoring. The 30/98, Vauxhall's most illustrious car, was built in various forms until 1926, scoring 75 'firsts' between 1920 and 1923 in sporting events, and was certainly the finest fast tourer made in Britain in the Twenties, marred only by its somewhat feeble brakes. It is still acclaimed as one of the world's greatest all-round cars.

At the time of the Prince Henry's success another sportive model in the luxury class had become a byword in Europe. Swiss engineer Marc Birkigt made the first of his legendary Hispano-Suiza (Spanish-Swiss) cars in 1904 in Barcelona, Spain. His superbly engineered models, designed for the carriage trade, were popular in the royal household of Spain. The King, young Alfonso XIII, had two in his own motor stables.

In 1910 a smaller version of the marque won a light car race, and Birkigt based a production sports car on it. The Queen of Spain presented one to her husband that year, and the car became the Alfonso (Type 15T) model of the marque. Birkigt's new engine, one that was to prove significant throughout the automotive world, was enlarged to 3·6-litres, which gave 64 bhp at 2300 rpm. Birkigt's engine of 1912 (with ohc) was sold (history is a little vague about whether it was sold or pirated) to designer Ernest Henry who modified it into his four-cylinder, 16-valve, twin-overhead-camshaft unit that was to greatly advance engine efficiency and prove a watershed of racing engine design.

Both the design and the materials used in the Birkigt units played a vital part in aviation progress during the First World War, when the Swiss engineer designed and produced a light, water-cooled, V-8 engine that enabled French aircraft to outfly the Mercedes-powered German

Bugatti, 2·3 supercharged Type 35B. More than 1,800 wins were scored during the 1920s by Type 35 Bugattis.

aircraft during the earlier part of the war.

The world of motor sport was by now truly international. In America, permanent motor racing venues had long since brought the state of the sport up to European standards. Henry Ford had earlier made and broken world records in his extraordinary '999' racer, and Barney Oldfield had become a household name on both sides of the Atlantic with his exploits in the Blitzen Benz, when he took it at speeds of over 2 miles a minute in 1910. Indianapolis had been opened in 1909 and the first of the famous 500-mile races had taken place in 1911. Brooklands in England had opened two years earlier (1907), had seen several categories of racing, a 24-hour endurance demonstration with an average of 66 mph, had matched Britain's best – a Napier – against the might of the Fiat stable for a wager, and had become the heart of the sport in Britain.

THE TWENTIES AND THIRTIES

The First World War pushed technical knowledge forward rapidly. When the dust had settled, motor sport opened for business with the 1919 Targa Florio and the Indianapolis 500 where the Duesenberg straight-8 was seen in competition for the first time. For the next few years Duesenbergs and Millers were the doyens of the Indianapolis track, and from 1922 to 1936 no other marque won there.

On the road there were several choices for the young American sportsman. The low-slung, slow-revving 60-hp Stutz Bearcat, with its bucket seats, token windscreen, and little else in the way of comfort, alternated with its rival, the Mercer Type 35, as favourites with the sporting man-about-town even though they'd been around since before the war. Upmarket, Duesenberg was making cars that, apart from clocking numerous racing victories, were being eagerly sought by the new rich of America, the film fraternity. Duesenberg's Model J of 1923 and its later development, the supercharged 320-bhp Model SJ, had performances that can hardly be believed today, including a top speed of 129 mph.

In the sporting world of Europe perhaps three cars captured the imagination of the crowds above all others during the 1920s – the British Bentley, the French Bugatti, and the Alfa Romeo from Italy.

RODO
5
5

W. O. Bentley had shown his first car, a 3-litre, four-cylinder model in 1919, and sold his first in 1921, when it won its first race at Brooklands. A 3-litre Bentley won the notorious 1927 Le Mans after it had been all but wrecked with its two team cars at White House Corner.

The four-cylinder, 4398-cc Bentley, made from 1928 to 1931, was fitted with a supercharger by racing driver Sir Henry Birkin (doubling its output to 250 bhp) and is perhaps the most distinguished Bentley of all. It won the 1928 Twenty-four Hours of Le Mans and later raised the Brooklands record to 137·96 mph.

Milan-born Ettore Bugatti made his first car way back in 1901, and won a medal at the Milan exhibition that year. He moved to Alsace, then a part of Germany, and worked first for De Dietrich, then for Emil Mathis, and finally for Deutz in Cologne. There, in the cellar of his own house it is said, he built his own *Pur Sang*, his thoroughbred automobile, first of a line (less than a total of 10,000) of jewel-like cars that placed Bugatti amongst the automotive immortals. This car, the 1327-cc, four-cylinder Bugatti Type 13, was developed during 1908–9, and Ettore moved to Molsheim in Alsace to begin his own commercial production. The car was an immediate success. Bugatti, ever restless, also designed a tiny 2-seater, 850-cc car which he sold to Peugeot in 1912. It became the popular Peugeot Bébé.

Le Patron, as Bugatti was respectfully called, produced model after model and by 1924 had built what was to become his most famous car, the racing Type 35, first seen in 8-cylinder, in-line, 2-litre form, Bugatti's classic layout. During the following years the Type 35 was seen with

Left: *The* Classic Grand Prix car? The 250F Maserati built from 1954 to 1957 was not remarkably powerful yet could often beat the most formidable opposition. In 1957 it took Juan Fangio to World Championship in 2½-litre form.

Above: German domination of Grand Prix racing in the Thirties led to revived interest in 1½-litre voiturette events. The ERA seen here was a successful contender.

Left: The Alfa Romeo P2 introduced in 1924 was a highly successful 8-cylinder, 2-litre supercharged Grand Prix car designed by Vittorio Jano.

supercharger (everyone else was fitting one so Ettore reluctantly had to follow suit); as a 35A with a different engine and crankshaft; as a 1½-litre car; as a special 1100-cc racer for the 1962 Alsatian Grand Prix; as the 2.3-litre 35T modified for the Targa Florio. The Type 35B, an improved 35T, also had a 2.3-litre power unit, supercharged, and the 35C had a 2-litre supercharged engine. Bugatti conjured up a different car for each type of race over the years 1924–31. The engines were architecturally square and always (in his Type 35 racing versions) of eight cylinders. Bugatti juggled with his engines to considerable effect, for in the hectic seasons between 1924 and 1927 his cars won an astonishing total of 1,851 races.

Modern rallying. A French Alpine driven by Italian entrants roars up a steep road in the hills of central France during a recent Monte Carlo Rally.

The 'tween-years were the heyday of open cars which had pretensions to some sort of sporting performance. Britain led the open-air vogue – and produced some of the most practical motor cars. Names like MG, Alvis, Lagonda, Frazer-Nash, Aston Martin, contested sporting events – and often completed illegally on the public roads – with France's Amilcar, Salmson, BNC, Morel and other examples of the legion of up-and-coming French makers. In fact, over 300 firms in France were producing cars during the early Twenties, all but three of them long vanished. The survivors are Renault, Peugeot and Citroën, who wisely did not place all their energies or their faith in the production of sporting vehicles.

Italy, quietly building its motor industry, had by 1921 seen the prototype of Vincenzo Lancia's new concept of car design, dubbed the Lambda. By 1923 it was in full production, this advanced design with its narrow-angle, V-type, four-cylinder engine and near-monocoque construction (using the strength of its body's rigid frame instead of a heavy separate chassis), and independent front suspension. It was basically a fast tourer but various private owners entered their Lambdas in sporting events and many a Lambda driver carried off the main prizes. Alfa Romeo, the other Italian classic sporting marque, had opted for more active participation in Grand Prix motor racing, and by 1924, after Vittorio Jano had joined the company as designer, the 8-cylinder, 2-litre

Juan Manuel Fangio, five times World Champion, takes the Mercedes-Benz W196 through a fast bend during the 1955 Grand Prix at Aintree in the North of England.

P2 appeared in the grid – and won the first Grand Prix it entered.

From 1934 the Mercedes and Auto Union companies of Germany were building cars that would conform with regulations which stated that racing cars should not exceed 750 kilogrammes. This limitation also gave birth to the famous Alfa Romeo Monoposto which in the hands of little Tazio Nuvolari ('brakes are no good . . . they just make you go slower') defeated the powerful German opposition on its own home circuit, the Nurburgring, in 1935.

The rear-engined Auto Union, in spite of its awkward handling, took just two seasons to beat the world. In 1936 the car, with former motor cyclist Bernd Rosemeyer at the wheel, won six Grands Prix, proving that efficient design could get tremendous power out of an engine that complied with lightweight regulations designed primarily to prevent engines becoming too powerful.

Mercedes-Benz had also been encouraged by Hitler's government to show the world the calibre of its product, which it did by building for the 1934 season the first of the classic Silver Arrow racing cars, the straight-eight, 354-bhp W25. When this was followed in 1937 by the immensely powerful W125 (with a 5.7-litre engine giving 646 bhp) the Grand Prix racing world was effectively tied up by the Auto Union and Mercedes-Benz, hardly any other marque recording a win that season.

One effect of the German Grand Prix victories of the 1930s was to revive interest in 1½-litre voiturette racing in other European countries. Britain, particularly, eagerly entered this class of the sport and the Riley-based ERA (English Racing Automobiles Ltd.) began to figure in the winners' lists in competition with Maserati and Alfa Romeo.

Meanwhile Mercedes-Benz had in 1939, under the 3-litre Formula, brought out the W163, a V-12 car with double stage supercharging in which one compressor fed another, finally giving the engine a mind-boggling boost to 480 bhp. One can hardly imagine what sort of car would have come next from the incredible Stuttgart stable if war had not halted the sport.

MERCEDES RETURN TO THE GRID

After a 15-year break, Mercedes-Benz of Germany continued with very much the same programme as before. By 1954 the Silver Arrows Mercedes-Benz Grand Prix cars were on the grid contesting the supremacy of the Ferrari and Maserati *équipes*. After a try-out with some over-streamlined cars, the Mercs began to start leading the field again, and by the end of 1955 it was all over. The Mercedes-Benz W196, straight-8, 2·5-litre, 270-bhp cars had captured all the available silverware once more. The classic 250F Maserati, which had during the first half of the decade mopped up most of the prize money in the hands of Stirling Moss and Juan Fangio (who were now driving for Mercedes), and the Ferrari which had taken others of the world's best drivers to victory, were totally eclipsed by the Mercedes W196.

Above: The British-built Ford GT 40 (40 inches high) first challenged the supremacy of the Ferraris in the Le Mans 24 Hours in 1964. It won in 1966, 1967, 1968 and 1969.

Right: Current Formula 1 (Grand Prix) regulations restrict engine size to 3 litres or 1½ litres supercharged, and a minimum weight of 500 kilos. This is a Brabham-Alfa Romeo on a very wet Nurburgring.

Happily for Grand Prix racing, the Stuttgart company felt that in two seasons they had made their mark on the sport and on the public and could withdraw from the Formula 1 scene with honour, having proved in no uncertain fashion that Germany was back in the race in more ways than purely sporting terms.

Like most other sports, motor sport has grown rapidly in popularity during the past generation. Encouraged by sponsors – commercial organizations whose cash payments defray the costs of putting a car on the start-line in return for advertisement and publicity – the sport has diversified into a score of classes and categories in racing, record breaking, hill-climbing, rallying and various other mud-plugging branches of competition.

In Britain the sport received several shots in the arm after years in the doldrums trailing behind German and Italian contenders when in 1958 Mike Hawthorn of Britain became World Champion, albeit in an Italian Ferrari. Then when the world title went to drivers such as Jim Clark, John Surtees, Graham Hill and Jackie Stewart in cars such as the British-made Lotus and BRM, Britain became the centre of motor sport for some years.

Today the sport has again begun to move its focus to other regions, to France for instance where our imagination is captured by turbo-charged cars like the Renault, and back to Italy, to firms such as Alfa Romeo again producing a racing engine. But the majority of teams, most of whom use Ford engines for their cars, and all of whom bring new cars to the grid every year, modified to clip another fraction of a second off last year's lap times, are very much a result of British technological

supremacy – and the undoubted genius of one man, Colin Chapman, designer and builder of the Lotus.

Today many regions of the world are providing drivers from outside Europe. The 1978 World Champion, Mario Andretti is from the U.S.A., (the first American to win the championship since Phil Hill in 1961). In 1979 the world title was won by South African Jody Scheckter. Gilles Villeneuve is from Canada, Riccardo Patresse from Italy, Daly and Kennedy from Ireland and Zunino from Argentina, drivers from whom future world titles will almost certainly be drawn.

The Golden Years

Early kit car. A series of articles in a magazine of 1900 offered to show readers how to build a small car. The result – a tidy model (this is a 1904 example) with a two-cylinder, 8 hp engine.

During the period between the years of experiment (generally accepted to be from 1885 to 1905) and the great changes that the First World War brought in its wake, motoring enjoyed what was then considered to be, by motorists at least, a healthy proliferation, although in Britain the anti-automobile crusade was at its peak. There the unfortunate motorist risked a fine every time he left his private garage. The police and the courts filled the local coffers with the riches extracted from the intrepid motorist for exceeding 20 mph, or allowing a lamp to blow out – in days when roads were so free and empty that one may have met perhaps a dozen other motor vehicles on the 50-mile journey from London to Brighton.

However, new makes and models from every manufacturing country

Dutch aristocrat of 1904, now resident in the Lips Museum, Drunen, Holland. This dignified pot-bellied model has a four-cylinder engine of about 18 hp. The car appeared in the film *Genevieve*.

were pouring on to a still-hungry market during the middle of the first decade of the twentieth century, covering the entire range of size, sophistication, costs and purpose, from flimsy tricycles with mouse-power motors to large and solid products such as Belgium's classic Minerva, or the de Dietrich from Lunéville in Lorraine. Even kit cars were seen in magazines of the day, some, like the English Mechanic, born as a series of instructional articles, features that when the going got too hard for the reader (or the author) switched to advice on where to buy the requisite parts. The English Mechanic was in fact one of the better kits and several have weathered the fortunes of the past 80 years to delight us today.

By the middle of the Edwardian period every manufacturing country had its firmly-established top marques. France, a leading motoring nation, had the Delaunay-Belleville, round and dignified and ponderous, followed by several other worthies such as Mors, De Dietrich, and the twin originals, Panhard and Peugeot. The lesser breeds, Bollée, De Dion, Ader, Unic (of taxi fame) and Renault were beginning to produce models for the masses. Germany's Mercedes reigned supreme on the other side of the Rhine, with Opel climbing fast in production numbers. Benz, still over 20 years away from amalgamation with Daimler, was fighting back after the triumph of the market-capturing Mercedes, and was winning a number of sporting events, while the little cycle firm of NSU had presented its first modestly-powered cars.

The Cadillac had become the undisputed quality auto of America by 1906, hustled by a gaggle of companies aiming hard for the same customers, with the three Ps (Packard, Peerless and Pierce-Arrow) as chief rivals. . .and of course Henry Ford's Model T was at this time right there on the drawing board, almost ready to take over the world's transport problems.

There is one car that embodies all that was best of the Edwardian age, the *belle époque* of motoring, one model of one marque that more than any other epitomizes the craftsmanship, the rapidly advancing engineering techniques, the atmosphere and the philosophy of the era before the First World War – the Rolls-Royce Silver Ghost. Called by its purchasers, by its admirers, even by some of its competitors, the 'Best Car in the World' (it was never called so by its makers), the Rolls-Royce captured the imagination and the enthusiasm of the small, wealthy motoring public as no automotive product ever had before.

Even the story of the making of the first Rolls-Royce cars and the subsequent meeting of the two men who were to form the famous company is typical of the attitude of an age when British workmanship was so often best: Royce, the meticulous engineer who would not accept

Another aristo, the Minerva, from Belgium. The 1909 model shown here was one of the first of the marque to use the Knight Sleeve valve engine, a silent and smooth unit used by several top makes.

anything less than perfection, and Rolls, whose agency for French cars left him unsatisfied with the quality of the product he was selling to the British public.

The first Royce car had been seen as early as 1904, an adaptation of a French Decauville that Henry Royce had brought over to his Manchester workshop, a 10-hp, two-cylinder machine with overhead inlet valves. The Hon. Charles Rolls heard of it – the only British car that could match and even improve on the continental makes – met the dour Royce, drafted an agreement, and formed the company that was to become the doyen of the automotive world.

After making several models – a 10-hp two-cylinder, a 15-hp three-cylinder and a 20-hp four-cylinder – in 1905 the new company made a 30-hp six-cylinder model. Then in 1906 the company laid down a one-model policy, and a brand-new 40/50-hp, six-cylinder, 7046-cc model was shown to the public at the Motor Show, clad in a silver-aluminium body. The Silver Ghost was born. It was made, with very few changes in basic design, for 19 years, and was bought by kings and presidents, by the famous and the rich from all corners of the globe, acquired for its prestige and loved for its grace, its silence and its trouble-free longevity. Today, Rolls-Royce will inform visitors who often enquire where they house their product museum, that there is *no* Rolls-Royce museum. All their cars are still in working order!

Until 1907, when the first hint of recession was seen on the horizon, the motor industry had plunged onwards to what it firmly believed would be everlasting prosperity and progress, an age of freedom and mobility for everyone. When for the first time the sales figures for the industry sagged, few realized that for a time at least production had exceeded demand. It was the first of the motor industry's regular recessions, and led to much re-thinking and re-grouping.

A note from a European motoring journal in 1907 makes curiously modern reading: 'A few years ago everybody wanted cars, regardless of cost, and with continuous fresh improvements; many took to buying cars every year, selling off the old.

Above: Stanley Steamer in commercial guise, made in Newton, Massachusetts, U.S.A. Steam power was favoured in that country by many in the early twentieth century. The last Stanley was made in 1927.

Top right: This 1905 Daimler 'Detachable' had hooks on top of its roof so that it could be winched off, converting the car to an open tourer.

Right: Standard 1907, when the word meant that the car was made to a high standard, not built down to a price! This fine 30 hp car has luxury *Roi des Belges* coachwork.

Below: The Rol s-Royce Silver Ghost was first seen at the 1906 Motor Show in London. The 40/50 hp Ghost had a 7-litre engine and a 4-speed overdrive gearbox, and was made for 19 years. The car seen here – the original model – is still in excellent working order.

Built by Napier in 1908 to the orders of a racing driver, this four-cylinder model was called a Hutton (Napier were promoting their six-cylinder cars at the time). The car created several Brooklands speed records.

'This golden age could not last. Customers soon discovered that the annual cost of a car was not only excessive but unnecessary and there was a general tendency to prefer comfort to speed. There was a considerable falling off in orders, and as the trade has gone on manufacturing, there is at present heavy surplus production. . . .' and the article goes on to say that numbers of French companies would succumb, and that their demise would consolidate the industry. We have heard similar sentiments many times since those relatively carefree Edwardian days.

In Britain the motor industry had centred mainly around the Midlands – Coventry, Birmingham, Wolverhampton – and workers had flocked to the region as they had flocked to the Lancashire mills a century earlier. Companies were formed, like Standard (1903), Singer (1905) and Hillman (1907) that were to offer solid and reliable material to the growing hordes of would-be owners, companies that would supply two generations of family motorists. On the luxury scale, Napier had firmly established their 'Six', largely by competition success, and of course Rolls-Royce was capturing the aristocratic market with Henry's Ghost, clothed in majesty by Barker, Mulliner and other eminent coach-builders.

Middle-of-the-road Riley were turning out 2-litre V-twins which looked a little like the round-radiator classic of France, the Delaunay-Belleville; only a year earlier in 1907, the Coventry company had been making crude tri-cars, three-wheelers not far removed from motorcycle design. Alldays were another with a product a cut above the popular markets, as were of course Daimler (now firmly wedded to their Royal clients, the British Monarchy) and the social-climbing Rover.

By 1910 cyclecar fever had hit Britain – and indeed all Europe – and many a motorcycle-engined, wire-and-bobbin bolide was seen weaving along the highway to the hazard of occupants and watchers alike.

Probably influenced by the light-car racing that was beginning to attract larger audiences, and certainly influenced by the desire of the lower-incomed man-in-the-street to have some sort of a four-wheeled motorcar – any sort as long as he could boast family mobility – shoals of cyclecars were pouring on to the unsuspecting market. To be fair, some were made with a degree of care, particularly the British three-wheeler made by H. F. S. Morgan in Malvern, Worcestershire. His first was powered by a V-twin unit from an abandoned motorcycle which he put into a light, tubular, three-wheeled chassis. Today the small factory at Malvern still painstakingly produces Morgans, now classic sportscars built in traditional style.

Another well-known pioneer of this dubious era was the Paris-made Bedelia, which, steered by a centre-pivot system and belt driven, looked about as roadworthy as a speedboat with its two seats arranged in tandem, one in front of the other. It suited the young, impecunious French sportsman however, and sold in large numbers. By 1912 over 40 different makes of cyclecar were shown at the Olympia Motor Show, ranging from the one-seater Rollo at 75 guineas to the tiny spidery GN which in time became the highly sporting Frazer-Nash.

From the cyclecar episode of motoring history came the light cars of immediate pre-war days. William Morris in England offered his first Morris-Oxford light car in 1913, and the little 10-hp, 1018-cc two-seater pioneered the so-called 'New Motoring's' graduation from the cyclecar era into respectability. Others joined Morris (as rivals rather than collaborators), in making cars that were genuine four-wheel pieces of engineering, rather than motorcycles with ideas above their station. Slightly larger than the first Morris was the Singer, a 1096-cc light car that had its gearbox housed in its back axle – and which inspired young Lionel Martin, who was looking for a car that he could tune up to high performance for competition, to build his first car in 1913, the car he dubbed Aston Martin. Peugeot of France offered the Bugatti-designed, 856-cc Bébé, another diminutive and simple car that exactly suited the new car-buying public.

The larger cars of the period were changing rapidly in appearance. Flatsided bodies and torpedo shapes (open four-seaters with a continuous line from the bonnet [hood] along the tops of the doors to the rear end) were becoming popular. On the other hand, weather protection was also being considered seriously for the first time. One foreign nobleman astounded the motoring circles by having the driving compartment of his landaulette enclosed by glass screens to protect his chauffeur from the

Alfa before Romeo. The first Alfas were made in Milan (at an old Darracq works) in 1910. This is the original model, a 4·1-litre, 24 hp open tourer.

elements, a design that was the fore-runner of the genuine *conduite intérieure*, as the French still tend to call a saloon or sedan. Tyres too began to play a part in protecting the passenger from the worst of the roads' ruts-and-bumps. Large-section tyres were used and pressures reduced in an effort to increase comfort. Electric lights were still very new, as were pressed steel wings [fenders], replacing the metal, wood and leather of yore, and pressed steel wheels.

The first practical self-starter was seen in 1912. American Charles Kettering, an electric motor expert working for Henry Leland's Cadillac Motor Car Company at Detroit, Michigan, had developed a motor that could generate enough energy over a short time to turn a heavy engine fast enough to spark it into life. The impact of this was greater than expected. Almost immediately a legion of new motorists appeared – lady drivers. Once the back-breaking necessity of turning the weighty and sometimes dangerous crank handle to start the engine had been overcome, the way was open for the woman at the wheel. The electric self-starter and lighting system became a standard fitting to the 5½-litre, 60-mph Cadillac of 1912, bringing about new body styling (elegance to suit a lady purchaser), new colours, a new advertising approach, and not least a new freedom for women, who could now undertake motoring journeys on their own without being considered indelicate or 'fast'. Before the following year was out most U.S. models were fitted with self-starters, although European manufacturers were slower to appreciate the advantages. Many of their clients already had self-starters. . .called chauffeurs.

THE AMAZING T

The greatest impact in the history of the automobile was made by the Ford phenomenon in the years before the First World War. Henry had started the Ford Motor Company in 1903 after a couple of false starts and was, by 1907, standing on the brink of the greatest development in motoring, his Model T, which was scheduled to be offered to an eager public in October 1908. 'Homely as a burro and useful as a pair of shoes' ran the advertisement, and they flocked for their new mechanical burro, buying more than 10,000 of them in the first production year – and continuing to buy the Model T until sales figures reached a fantastic 2 million a year during 1923–5.

The power for this little workhorse was a 20-hp, four-cylinder, three-bearing unit cast with an integral water-jacket and detachable cylinder-

Below: Riley of Britain made this 1907 V-twin two-seater, their first genuine car. Earlier the Coventry firm had produced tri-cars but in 1904 had moved to four-wheelers.

Right: Rover had made pedal cycles, progressed to motor cycles, and in 1904 made their first car, an 8 hp single-cylinder model. This is a 1907 example.

DUNLOP CORD MOTORCYCLE EXTRA HEAVY
Metro-Dodwell Motors

head. Its somewhat tricky planetary transmission had just two forward speeds and was designed to prevent too much damage being done to the gears by ham-fisted farm boys from Ploughsville. Capable of 45 mph and about 20 miles to the gallon, the Model T, the loved and derided 'Tin Lizzie', rapidly became part of the American scene in the days before the First World War.

Used in the city by salesmen and doctors (the first to appreciate its value) and down on the farm or ranch by country folk, the Model T offered not only utilitarian transport but could be adapted for sawing logs, hauling loads out of ditches, even milking cows. And when Ford moved into a new 60-acre factory and introduced modern mass production methods, the Model T was way out in front of any other motor maker in the world, using one-sixth of the nation's automotive labour force and producing one-half of the American nation's cars. Ford stock made fortunes for the lucky investors who had placed their faith and their

cash in the company. James Couzens, for instance, had put up $100 when the company started. He drew out $95,000 in dividends and in 1919 sold his shares for no less than $260,000.

While Henry Ford was building his automotive empire, W. C. Durant was playing out a similar drama with a number of other motor companies in the United States. The difference was that Durant was an entrepreneur, an ace salesman, rather than an engineer, although he had earlier built buggies for the U.S. market. His undoubted talents lay in persuasion and in organization. But his great energy, or his over-enthusiasm, eventually led him into trouble. Guided by Durant, the companies of Buick, Oldsmobile, Oakland (now Pontiac) and Cadillac formed the nucleus of what was to be called General Motors in 1908, a conglomerate designed to produce automobiles that would satisfy all

Below left: The sportive Hispano-Suiza Alfonso of 1912. Although the first cars were made in Spain from 1904, the company's Paris assembly plant sold well and the marque became thought of as a French product. Later, cars were manufactured in Paris.

Left: The French company of Sizaire-Naudin made single-cylinder cars long after most makers had moved to multi-cylinder products. However, the car had introduced many technical innovations.

Above: Lanchester 1913. An English pioneer from the dawn of the industry, the marque was always ahead of its time. This car had a 36 hp six-cylinder engine and seated five.

sections of the market. He had originally tried to include Ford in his far-reaching plans but Henry proved too awkward a cuss for him and the deal fell through.

Durant bought everything in sight during the first couple of years of General Motors – accessory makers, electrical companies, parts manufacturers and a bag of smaller concerns like Elmore, Ranier, Rapid, Welsh and other sideliners. Durant built, in just two years, a group that looked impressive and indeed was a breathtaking example of empire-building, but there were too many rotten apples in his barrel and the new group descended into chaos. Henry Leland of the respected company of Cadillac bailed him out. Later, other names, some still familiar, were added to the General Motors stable – Chevrolet, Fisher Bodies, AC Spark Plugs, Frigidaire.

William Richard Morris had repaired cycles in the 1890s and started to design cars in 1911. This is his first, the Morris Oxford Bullnose of 1913.

MEANWHILE...EUROPE

The golden years of European motoring were brought to a sudden halt soon after a Duke was shot and killed. The car in which Archduke Franz Ferdinand of Austria and his wife met their end was a Gräf und Stift, a car of humble origins – its makers were former cycle repairers in Vienna – but which had become known, over the 20 years of its life, as the Rolls-Royce of Austria (a title which tends to confirm the British company's supremacy as much as anything else) and which was perforce one of the historic cars of its time.

But the half-dozen years leading up to that tragic event, the moment in history that can be pinpointed as the end of an era and the beginning of the conflict that changed all our pasts and our futures, produced a number of fine pieces of personal transport for the wealthy few, and an increasing number of more utilitarian vehicles for others who until then would not have known the pleasure of owning any private transport more elaborate than a pedal cycle.

In 1912 Britain showed the short-nosed, 38-hp, six-cylinder Lanchester at its motor show, now a regular annual event that was part of the motoring scene, and the already legendary Prince Henry Vauxhall (chassis only: £495). In 1913 a 10-hp, four-cylinder AC made its debut; the Surrey-based company had been making funny little hansom-cabby, three-wheeler 'sociables' until then. The same year also saw a car, subsequently made for almost ten years, that well illustrated the changes that had swiftly come about in the design of, and attitude to the new age of motoring, and pointed the way to the changes about to take place. . .the Sunbeam, an immensely successful touring car, designed by Breton engineer Louis Coatalen who had worked successfully for Humber and Hillman. A designer who built and raced the cars he dreamed up, Coatalen was a man for all seasons for his company and was the mainspring of Sunbeam's illustrious reputation of later years.

This 1919 'economy' Stellite was made by Wolseley. The car made its debut in 1913, using a four-cylinder, 1100 cc engine.

Other models – the large Napier saloon for example – showed how coachwork design had at last broken free of traditional horse-carriage influence, with its low lines and enclosed bodywork that provided complete cover for the driver as well as 'paying' passengers. On most cars one saw clean and simple electric lighting by 1914, and detachable wheels had cut out some of the frightful problems of changing tyres after the still-inevitable puncture. Brass was slowly disappearing and the trend towards using the car as an instrument of daily labour was more evident in those last pre-war days.

French designs had confirmed that the single cylinder was now dead (the Sizaire-Naudin had made a 'one-lunger' until 1910, stepping up power over the years by enlarging the diameter then the stroke of its lonely piston). A single (monoblock) casting for the cylinders was now normal, doing away with much of the copper spaghetti and nuts-and-bolts of earlier designs. Split-V screens were seen, and the overhanging saloon roof which acted as a sort of sunvisor for the driver was fashionable, as were open, square-cut, torpedo-tourer shapes. A Lancia Theta could be bought with a flowing moulded landaulette or sedan body with lines that belonged to a decade later; the Peugeot Bébé of the day was surprisingly similar in profile to Austin's Seven of nearly ten years later; Itala of Turin made a tourer that looked like the great tourers of the Twenties, as did the British Valveless, and the small French Sigma (a modest 'assembled' car) had the distinct appearance of post-war build.

Coal-shovel bonnets [hoods] abounded in France as Renault took over the lead in popular production, and many a wise competitor copied the unorthodox Renault front end. From diminutive Decauville to majestic Delaunay-Belleville (with its round-bellied radiator similar to the Dutch Spyker), protection from the weather had assumed paramount importance, and by 1914 large glass areas surrounded the passengers and chaffeur alike. Motoring clothing, up to then a vital part of the traveller's

Last of an era – a 1913 Sunbeam tourer. The British company had started life back in the nineteenth century enamelling tinware, had moved into bicycle making and thence to cars.

wardrobe – goatskin great-coats with the fur exposed or racoon coats, dust veils, masks and so on – were at last thrown into the attic to become sought-after relics half a century later.

The so-called Edwardian period was one of rapid advances of fundamentals during its early stages, and of increasingly sophisticated refinements during the last year before the First World War. The car was not yet a completely accepted method of transport for the work of the world, but progress in engine mechanics, in suspension, in bodywork and, most importantly, in manufacture, had brought the automobile to the brink of modern usage as a functional artifact with an essential role to play in everyday life.

The Years Between the Wars

If Herbert Austin had produced only his Seven of 1922 he would still have made history. The diminutive Austin Seven brought motoring to an eager post-First World War public.

The period immediately following the ending of hostilities in 1918 was a turbulent one in which the scramble to turn swords into ploughshares was frustrated by post-war shortages, inflation (the like of which has probably not been seen since) and labour difficulties.

A four-year war of attrition in which virtually every major nation in the world was involved – and many of the smaller ones too – with all the waste of raw materials which that involved, meant that many of the larger manufacturers simply could not obtain the steel and other commodities they required to maintain output at a rate which would make their war-inflated factories economic to run.

The cost of the war in terms of men and materials meant that the average British car of 1919 cost practically double the figure of its 1914 counterpart, and increases of this nature were reflected in all other spheres. Initially, this doubling in price did not deter potential buyers, for there was a desperate shortage of cars, but when the supply situation eased by the end of 1920, prices began to fall again.

Many of the larger manufacturers were still in a difficult position, however. Ordered into munitions initially, and producing nothing else for four years, many were surprised by the Armistice, and caught with no new models 'on the drawing board' and with many machine tools either unsuitable for peacetime production or worn out from constant use.

America's involvement in the war, although massive at the end, came later, and consequently she had expanded her markets considerably between 1914 and 1916 and up until her own entry into the war in 1917. She therefore had a three-year start, more up-to-date models and many new customers who had been wooed away from British, French and German manufacturers during the years that those countries had ceased manufacture for civilian consumption.

This then was the scene at the beginning of the post-war period, and at first, when the big battalions were re-converting their factories and generally girding their loins to meet post-war demand, many small newcomers hopefully joined their ranks. These firms offered an assortment of 'assembled' cars to a public who were desperate for transport of any kind. Many were poorly designed and constructed and most of them relied upon specialist outside suppliers for their engines, gearboxes, axles and other major components.

Below: Aircraft-maker Gabriel Voisin had a dislike of any design that was not strictly functional, a taste that showed itself in the strangely angular shape of his cars. This is a Voisin C3 tourer of 1923.

Right: One of the great Mercedes-Benz line of the Twenties, the 7·1-litre SSK, the short-chassis version of the SS, both of which, although called sports-tourers won Grand Prix events against pure racing vehicles. The SSK developed 225 bhp in supercharged form.

A steam-lorry maker once said that his vehicles lacked the 'thump' of his Bentley as it loped down the road – and Bugatti called them the fastest lorries ever made. But the Bentleys of the Twenties, often known by their labels, Green, Blue, Red, were greatly loved British sports cars.

In America particularly the assembled car enjoyed a longer life than across the Atlantic, and literally hundreds of firms mushroomed during the Twenties. On both sides of the water, however, while many of the newcomers had a history of general engineering and war work in munitions, as many again had no engineering background whatever. Canned soup, government surplus and vending machine distributors were just a few of those trying their hands (usually unsuccessfully) at motor manufacture during this period.

There were others though, particularly those who had been engaged in the aircraft industry – which had greatly expanded during the war – who turned their new technology to good advantage in peacetime. Salmson of Billancourt, France, for example, had been renowned for their water-cooled and later air-cooled radial aero engines, and in 1919 they entered the arena of motor manufacture by the simple expedient of taking out a manufacturing licence for the GN cyclecar, already being produced in England. This rapidly developed into a fully-fledged small sportscar of the type which dominated the French scene throughout the Twenties. Initially a rather unorthodox little design incorporating a differential-less rear axle and a four-cylinder engine with a single pushrod per cylinder which also operated the inlet valves as a *pull* rod! Salmson's first 'home-grown' design was the work of Emil Petit. He later developed it into a twin overhead cam unit of remarkable efficiency and speed, but still of only 1100-cc capacity and with a two-bearing crankshaft. In this form, a Salmson won the 1921 Cyclecar Grand Prix and took second place (ironically behind a British GN) in the Brooklands 220-mile Race.

As if all this was not enough, first and second places were taken in both events in 1922. The sporting character of the Salmson was assured and the zenith of its career was achieved in 1927 with second and third places

in the general classification at Le Mans.

Touring model Salmsons acquired the overhead camshaft engine in 1922 but towards the end of the decade the small sports car vogue in France disappeared virtually overnight, and the arrival of the cheap MG Midget on the British market killed what British sales might have kept the type alive after 1929.

THE LUXURY MARKET

Also located at Billancourt, the Farman brothers had made their reputation with aero engines, and the design of the 1920 6·6-litre A6B Farman reflected this background. Much use was made of aluminium alloy in the construction of the engine, valves were actuated by a single overhead camshaft, and the cars boasted a peculiar suspension system combining a single transverse spring with two cantilevers.

Apart from the upstairs camshaft, Farmans bore no resemblance to their neighbours, Salmson, and were designed wholly for the luxury market. Whereas Salmson's licence-produced GN had sold some 3,000 examples by 1921, Farman's total production up to 1931 was a less than impressive 120, and it is remarkable that any should have survived from such a small output. In fact, about four or five are preserved today.

Built in exactly the same idiom and aiming at an identical market, the H6B Hispano-Suiza announced in 1919 at the Paris Salon was the first French-developed Hispano model and, at first, no faster than the Farman. However, it was hailed as the sensation of the show, and boasting four wheel brakes five years before Rolls-Royce grudgingly gave them to the Silver Ghost, the car was equally at home with sporting bodywork or bespoke formal coachwork, and was powered by an engine identical in capacity to that of the Farman.

The Vauxhall 30/98 (it was never discovered why the car was given this name) was the company's most successful sports car. Produced until 1926 it was used extensively as a competition vehicle, but was flexible enough to be used as a family car.

Often called the 'Baby Lincoln' – it looked a little like its more expensive cousin – the 200 cu in Model A Ford, introduced in 1928, was the successor of the Model T.

The Hispano-Suiza's fixed cylinderhead was a somewhat archaic feature, but otherwise the design was advanced, being one half of a V-12 aero engine developed during the war but never used. With a seven-bearing, pressure-fed crankshaft and a single overhead camshaft operating two valves per cylinder, the engine gave 135 bhp at 3,000 rpm, combining a good power-to-weight ratio with first class reliability in the best traditions of aero practice.

So flexible was the engine that it could be accelerated from 10 km/h to 80 km/h (6 mph to 50 mph) *in top gear* in only 21 seconds, a feature aided by superb performance and handling. Admittedly it was not quite so quiet in operation as other luxury marques – a fact which probably contributed to a preponderance of sports models having survived – but its success with the 'carriage trade' was sufficient to give Rolls-Royce a severe shaking.

Down at Issy-les-Moulineaux, pioneer aviator Gabriel Voisin had built some 10,000 aircraft, mainly for the military, by the time the Armistice was signed, and he entered the motor industry in much the same way as Salmson had done, by taking over someone else's design. In Voisin's case it was to André Citroën that he turned, taking over the manufacturing rights of a Citroën design which Citroën himself had never used. The 18-hp, four-cylinder utilized the double sleeve-valve Knight 4-litre engine and was produced as the C1 Voisin. Henceforward, Voisin became wedded to sleeve-valve engines, and consistently developed them to a high degree.

The Voisin was a first class motorcar, capable eventually in C1 form of 130 km/h (80 mph), and numbered many titled owners among its customers. Voisin bodywork frequently combined aviation construction techniques with Voisin's own eccentricities to produce a strange mixture of practicality and uncompromising ugliness, and doubtless some of his more bizarre designs contributed to customer resistance.

BENTLEY DAYS

Such was the spate of newcomers to the manufacturing scene that the Olympia Motor Shows of both 1919 and 1920 overspilled to the White City Stadium, and relegated to this overspill in 1920, three newcomers shared adjoining stands. They were representative of the three major

Ford halted production of the T in May 1927, just after the 15-millionth example had rolled out of the works. This Tudor (2-door) one of the last of the line.

A French car with a sporting and quality reputation, the Delage was produced from 1905 to 1954. This is a 1927-8 2·1-litre Delage DIS/6 tourer.

categories of new make to be found on both sides of the Atlantic during this period; one of the three was destined to achieve immortality, the other two, obscurity.

Stand 426 housed Bentley Motors of Cricklewood, who had already shown their new 3-litre model at the 1919 show. Famous for his wartime aero engines, 'W. O.' Bentley had pre-war experience of the motor trade importing DFP cars (and competing with them) and the Bentley radiator owed more than a passing resemblance to the DFP.

Like the Hispano, the 3-litre Bentley relied upon a fixed head and single overhead camshaft and developed about 70 hp in its early guise. Only 1,600 were made, and yet the whole concept of motoring in the Twenties conjures up a picture of an open-bodied Bentley roaring open-throated down a tree-lined, straight, unmetalled road, driven by a University-scarved young man accompanied by a slip of a girl with an Eton-crop hairstyle and a cloche hat.

If the small sports car was the exclusive province of the French until the advent of the MG Midget, then the long stroke fast tourer represented the British scene, and the Bentley was the epitome of the fast tourer. Other makes emulated it, like the HE from Reading, and there were Continental counterparts in the Type 51 Itala (which distinguished itself in competition from the Targa Florio in Italy to the Aspendale Track in Australia) and the D1 Delage, at 2·1-litres smaller than the Bentley but in sports form utilizing the aluminium pistons for which Bentley was also famous.

Offered in short and long chassis form as the Red Label and Blue Label respectively, the Bentley 3-litre was supplemented in 1926 by the 6½-litre, six-cylinder model, the intention being to capture a proportion of the available carriage trade. The attempt was unsuccessful but not wasted, since it resulted in the 180 hp 'Speed Six' of 1929 which is considered by many to be the best model Bentley ever designed, and which was responsible for the last two Le Mans wins – Barnato/Clement in 1929 and Barnato/Kidston in 1930.

The 3-litre Bentley had developed into the 4½-litre by 1927, still a four-cylinder model but with sufficient potential for its 100 hp to be increased to 130 hp before production ended, and won Le Mans in the hands of Barnato and Rubin in 1928. Fielded in supercharged form (of which 'W. O.' himself did not wholly approve) by Sir Henry 'Tim' Birkin and

developing 182 hp, a 4½-litre took second place in the *formule libre* French Grand Prix of 1930.

Bentley's final significant car was the superb 8-litre six. Only 100 were built before Bentley went bankrupt in the summer of 1931. Had it not been for this last model (which presented a considerable threat to Rolls-Royce at the top end of the market) it is debatable if Rolls-Royce would have bid for the company.

THE EVERYMAN CLASSICS

Well-known motorcycle makers Clyno of Wolverhampton started later in the day than most, in 1922, and produced an excellent 'assembled' car for Mr. Everyman which ran William Morris and his 'Bullnosed' Cowleys and Oxfords a close second for a few seasons. Morris was the more

An American with a distinguished pedigree, this 1933 Packard V12 convertible-Victoria Model 1006 had bodywork by Dietrich (Owner A.F. Mirremiar).

successful of the two however, and was the first manufacturer to break the 40,000 units a year barrier. Unfortunately Clyno elected to enter into a price war with Morris, sacrificed quality for quantity, and lost the battle.

Quite apart from the small-fry, however, the established manufacturers did not all have a smooth ride in post-war years. Austin, in business on his own account since 1906, made a bad mistake in concentrating all his post-war production energies initially upon a single model, the Twenty. Excellent car it may have been, but it was nudging the luxury class, and with post-war inflation, too expensive for the mass market for which Austin's factories were geared.

A versatile engineer, Austin introduced the Heavy Twelve as a supplementary model in 1921. Initially with a 1661-cc side-valve engine and a later enlarged to 1861-cc, this model proved virtually indestructible

Below: The '20', introduced in 1922, was a considerable improvement on the ageing Ghost. This is a 1926 20 Hooper-bodied Landaulette.

and remained in production in taxicab form into the late Forties. It contributed as much, if not more, to the post-war recovery of Austin as did the famous little Austin Seven.

The brainchild of Herbert Austin himself, the Seven was conceived at a time when the fortunes of the parent company were at a low ebb and, so it is said, Herbert – known universally in the works as 'Pa' – would disappear for long solo runs on his bicycle rather than face the bank on paydays! As his Board of Directors was unimpressed with his plans for a new baby car, Austin borrowed the services of a young draughtsman and set up a drawing office in the billiard room of his own home.

The result was the Seven, announced to the Press in 1922, and intended primarily to appeal to the family man who wanted something better than a motorcycle combination, did not fancy a crude cyclecar, and could not afford a full size motor car. Like Bugatti's Peugeot Bébé of a decade earlier, the Austin Seven was in effect a large car in miniature, and combined all the virtues of a large car – water-cooled, four-cylinder engine, gearbox and live axle – with an economy of operation and low price. It was an immediate best-seller, and continued to find an appreciative market for the following 17 years.

Still at the popular end of the market, the dominant force during the Twenties was Ford, firmly entrenched at River Rouge in Detroit, and with an assembly plant at Trafford Park in Manchester, England dating from 1911, with various other assembly operations throughout the world. His staple offering (in fact, the only model available) since 1908 was the Model T, an extremely cheap, husky, four-cylinder car with a reputation for reliability with minimal maintenance which had passed into legend.

Beloved by Henry Ford and cursed by two generations of motorists for those aspects of its design which Henry steadfastly refused to update, the Model T soldiered on until 1927, by which time it was running a poor second to Chevrolet in the sales race. Finally Ford characteristically shut down the whole plant while a new model was prepared, and after seven months of apparent inactivity, rumours of impending bankruptcy were rife.

The new car appeared in 1928. The Model A was a solidly-built workhorse with pleasing lines and a radiator design borrowed from the luxury Lincoln which Ford had purchased when Henry Leland ran into financial difficulties in 1922. Over 633,000 were sold in 1928, and 1,507,132 Model As found customers in 1929, re-establishing Ford's lead over Chevrolet.

Chevrolet themselves had started modestly in 1911 when, together with racing driver Louis Chevrolet, William Durant brought out a 4·9-litre with side valves in a T-head. This was followed in 1916 by the '490' (the name was taken straight off the price tag), a four-cylinder overhead valve four costing only $490. In 1917 sales totalled a healthy 70,701 units and in that year General Motors acquired the company. Billy Durant went on to form yet another empire as Durant Motors, and this subsequently embraced Locomobile, Flint, Rugby and Star as well as his low-priced and initially best selling Durant Four.

U.S. LUXURY PRODUCTS

Spiritual home of the mass-produced car America may have been, but it would be quite wrong to imagine that she relied totally upon foreign

Bottom: One of the classic cars of the Auburn-Cord-Duesenberg Museum in Auburn, Indiana, U.S.A., is this 1935 Auburn boat-tail Speedster. The 150 bhp models were guaranteed to have been tested up to 100 mph.

Below: Alfa Romeo 2·6 litres, a sporting aristocrat from Milan. Vittorio Jano's 8C 2300 supercharged Alfa Romeo had appeared in 1931 and had won four Le Mans in a row, and three Mille Miglias.

imports to satisfy home demand for luxury vehicles. Although popularly associated with cheap Fords, Dodges, Chevrolets and Overlands – chiefly because these constituted the bulk of American exports to the rest of the world – America espoused some of the best built classic designs to grace either road or race track.

At Indianapolis, home of the hallowed brick racetrack, Fred and Augie Duesenberg at first built racing cars – their Bugatti-inspired 3-litre straight-eight with overhead camshaft and three vertical valves won the French Grand Prix in 1921 and their first production car was unveiled at the end of 1920. Generally accepted as being the first American production straight-eight, the Duesenberg Model A was not only extremely advanced, it was also decidely expensive.

Below: Aston Martin's 1½-litre long-chassis model had a classic front end – in this case well covered with stone guards, common in the days of rougher roads.

Generally acknowledged, together with Peerless and Pierce-Arrow, to be the Rolls-Royce of America, Packard were fielding a twin-six (the world's first series production car with this configuration) in 1920, as their sole offering, but this was in turn joined by a side valve single-six

which remained in production until 1928 and which was responsible for the majority of Packard sales.

Above: The Duesenburg Model J had an eight-cylinder, 6·9-litre Lycoming engine that developed 265 bhp; double that of its nearest rival. Seen here is a 1929 Dual-cowl Phaeton.

Connersville, Indiana was the stamping ground of the luxury McFarlan, the reputation of which has grown out of all proportion to the number of cars actually sold. At most, 235 cars a year were produced (and frequently less), yet the McFarlan was typical of all that was best among buyers of bespoke motorcars. The TV (Twin-Valve) Model utilized a massive engine embodying no less than three ignition systems involving the use of 18 sparking plugs, and at $9,000 the Knickerbocker Cabriolet was one of the most expensive American cars of the period.

Oakland, California however, enjoyed the distinction of being the home of one of the largest – and almost certainly one of the most expensive – American makes. At 13½-litres and giving 125 hp, the Fageol was priced from $9,500 (for a bare chassis) to $17,000 complete.

It would not be fair to deal with America's luxury makes without some mention of the American Rolls-Royce, although the Springfield, Massachusetts-built cars were never totally successful in the commercial sense. Unlike their British counterparts, they were offered complete with bodywork if the customer so wished. In addition, custom bodies were built by outside firms including the established and respected house of Brewster, whose bodies can be found on many of the 1703 examples of

Silver Ghosts and 1241 Phantom Is built in the American plant.

As in pioneer days the development of the American car during the Twenties did not follow exactly the same pattern as Europe, and just as the high-wheeler was a typically American phenomenon in the mid-Edwardian period, so the persistence of the steam car – with a mild revival in 1923–4 – was peculiarly American. Even the old-fashioned Stanley managed sales of some 600 per year up until 1925 (the last cars were made in 1927), and perfectionist Abner Doble was offering his advanced steamers until 1931.

Still leading the luxury field at home in England, and despite the inroads made into the bespoke field by Hispano-Suiza, Isotta-Fraschini and the American Cadillac, Rolls-Royce persisted with the now ageing Silver Ghost – still basically a 1906 design. Then in 1922, the Ascot racing fraternity, picnicking with a Fortnum hamper and a bottle of Bollinger, were appalled to find themselves confronted by a 'Baby' Rolls – The Twenty. Rolls had produced an economy car!

Below: *Traction Avant* was the French name for it – the 1934 front-wheel drive 7CV Citröen that was years in advance of most cars of its day.

Beloved of nouveau riche and war profiteers (*Punch* called them post-war sportsmen), the Twenty was frowned upon by the Establishment when first announced, in the event quite unfairly. None of the standards with which the larger car had become associated were sacrificed in the smaller model, and with its monobloc engine, overhead valves and central gearchange it was a decided improvement over the Ghost in many respects. The arrival of the New Phantom (Phantom I) for the 1926 season silenced the critics, although many felt that no worthy successor to the Ghost had been found until the advent of the Phantom II in 1929.

From the mid-Twenties the wind of change had been blowing strongly, and the first casualties, no matter how long-established they may have been, were those who had consistently ignored the warning signals. The First World War had swept away half the crowned heads of Europe, and with them the wealth and estates of many of the old aristocratic families who had made up their courts. These were the traditional buyers of, at first, horsedrawn carriages and, later, luxury motorcars, and their disappearance struck many a mortal blow.

The celebrated Delaunay-Belleville, that round-radiatored symbol of Edwardian opulence from St. Denis, Paris, the car which had been dubbed the Gallic Rolls-Royce, was never the same after the fall of Russia to the Bolsheviks (the Tzar had kept a stable of Delaunays), and whilst

The Austro-Daimler of Vienna was originally an offspring of the German Daimler company. The ADR8 was its only eight-cylinder car, built between 1930 and 1933.

they survived the Depression and staggered on through the Thirties utilizing American and German engines, the malady which beset them dated from the First World War.

Many other old-established companies were similarly afflicted, particularly in Belgium, once the home of proud makers like Excelsior and Pipe. Both of these succumbed in 1932, and even FN survived only on the strength of their ordnance division during the rearmament of the Thirties.

Towards the end of the Twenties there was a growing demand for cheap saloon-bodied cars. Mr. Everyman was tired of struggling in the pouring rain with one-man hoods which needed six hands, and all-weather equipment which wasn't – and in America Dodge, with their all-steel saloon, and Essex with their enclosed 'Coach' had shown the way.

The day of the composite bodybuilder was on its way out, and makers of cheaper cars were finding that they had to be comfortable, reliable and easy to operate if they were to remain competitive. Inevitably rationalization won the day, and despite his worries about unemployment, the average family man in the early Thirties obtained a far better deal when he bought a new car than ever before.

William Morris led the way with his £100 Morris Minor in 1929, and although his open two-seater with its rather complicated overhead camshaft engine was not too successful, it implanted the idea of a £100 car in everyone's head, and it was Ford, pioneer of cost-cutting whilst maintaining quality, who offered the first £100 saloon, the Y type two-door Popular, in 1935.

By this time, Ford's British Trafford Park assembly plant had given way to a purpose-built factory on reclaimed marshland at Dagenham in Essex, opened in 1932, and the Y type was not only the first truly British Ford, it was absolutely right for the economic climate of the day.

Austin's Seven, updated as the streamlined Ruby, was still giving good service and keeping the factory fully employed; it was also offered, in two-seater form, at £100. For those who wanted something a little larger, the Ten introduced in 1932 was a best-seller. Morris hit the jackpot in 1935 with his Series One 'Eight'. The accent of the day was on small, cheap cars, and even relatively small manufacturers like Jowett and Standard – the latter with their worm drive Nine – were able to weather the worst of the storm.

Most stylish American casualty of all was, perhaps, the mini-empire of Erret Lobban Cord, an entrepreneur *par excellence*, who having acquired control of Auburn – an old-established company – in 1924, proceeded to revamp its cars completely. Duesenberg came into the Cord fold in 1926, and Fred Duesenberg's first car for his new employer (Cord had stipulated it had to be new and quite exceptional) was the Model 'J' announced in 1928.

In looks, refinement, size, specification and speed it surpassed anything at that time being offered by other American manufacturers, and its Lycoming engine (Cord also owned Lycoming), a 6·9-litre eight and twin overhead camshafts and four valves per cylinder (quite unique for an American design) produced no less than 265 bhp, more than double its nearest rival.

In 1929 Cord optimistically produced a car under his own name. Designated the L-29, it was the first front-wheel-drive car to achieve anything like public acclaim and when it reappeared in redesigned form with a coffin-nosed body and disappearing headlights from the drawing board of Gordon Buehrig as the Model 810, it established a classic line which has since been acclaimed by bodies as far divorced from motoring

Above: The SS Jaguar of 1938 had a 3·5-litre ohv engine developing 125 bhp, an attractive and inexpensive sports-car of the Thirties, and direct ancestor of today's XJS.

Left: The coffin-nose body and vanishing headlamps of the Cord 810 and 812 of the mid-Thirties plus its vivid 4·7-litre V8 (supercharged in this 812) made it a memorable car.

Vincenzo Lancia's Lambda appeared in 1922, introducing new methods of construction. This is a 1929 S1 model (somewhat modified) that once belonged to the Italian dictator Mussolini.

as the Design Council and the Museum of Modern Art.

The demise of the Cord empire in 1937 (when the worst of the Depression was over) remains one of the ironies of the period. Cord's elegant and advanced cars enlivened those dark days, and have deservedly achieved the stature of all-time classics.

Cord was not alone in espousing front-wheel-drive, however. André Citroën, most successful of all the newcomers in 1919, popularized it with his 7CV 'Traction Avant' in 1934. A design demanding a completely new factory, the car was an instant success and remained in production until succeeded by the equally iconoclastic DS19 in 1955. The expense of the factory and retooling forced Citroën to sell out to Michelin, however. Such is the penalty of being ahead of one's time.

Although the British Alvis company's venture into front-wheel-drive was abortive, they made a most successful transition from the out-and-out sports car of the Twenties to the fast and stylish drophead coupes, tourers and sporting saloons which characterized their output during the Thirties. The Speed Twenty, and the later Speed Twentyfive were good looking, fast enough, comfortable, and handled and stopped well. In these attributes they echoed their contemporaries – the Derby-built 3½- and 4½-litre Bentleys, the flexible and low-slung Meadows-powered Invicta, the SS Jaguars, forerunners of today's XJS, and even the large and luxurious Austro-Daimler.

For the traditionalist, sports car *aficionados* could still buy a chain-drive Frazer-Nash up until 1939 (although the company were importing BMWs from 1934 onwards), or a three-wheeler Morgan (latterly with a Ford four-cylinder engine). The Thirties also heralded the arrival of the synchromesh gearbox, a refinement and ease of operation never dreamed of by motorists in 1919.

In Italy and Germany, Fascist governments saw, in Alfa-Romeo and Mercedes-Benz respectively, the means whereby nationalistic fervour could be encouraged through the medium of international motor sport, and both factories, together with Auto Union, received massive subsidies. In the case of Mercedes and Auto Union, these had the effect of making the German teams virtually unbeatable for a period, but in fairness, both Alfa and Mercedes had a respectable sporting pedigree, and the SS and SSK Mercedes series of the late Twenties and early Thirties are now acknowledged as some of the greatest sports car designs ever built.

Both Germany and Italy contributed a great deal to the advancement of the design of ordinary cars during the two decades, Lancia's Lambda pre-dating most of its contemporaries by more than a decade, and Germany absolutely bursting with innovation under the influence of engineers of the calibre of Porsche and Ganz. Independent suspension, development of two-stroke designs for motorcars, rear engines – all saw the light of day on a popular scale during the ten years leading up to the Second World War.

The twenty years 'between the wars' were robust, ever-changing, vital years in the development of the automobile, and the level of technology concerned with that development determined that the Second World War would be almost totally mechanized.

The New Generation (1950-1980)

British sporting classics. A 1973 1275 cc MG Midget and an MGB, with 1798 cc under the bonnet [hood], rest briefly in idyllic scenery.

The effects of the Second World War on the automobile manufacturers of the world were, in many ways, similar to those resulting from the 1914–18 conflict. As before, all had been diverted to war work, and similarly the rigours of war tested designs to the full, resulting in rapid developments and improvements. But there the similarity ceased.

Whilst the motor vehicle – and the aeroplane – had played an important part in the First World War, the Second World War was a war of far greater mobility both on the ground and in the air, and the growth of air power in particular was to replace, to a great extent, the concentration upon shells and other projectiles which had engaged so many manufacturers during 1914–18.

Civilian motor manufacture, which had been maintained at a trickle throughout the First World War, halted in Britain altogether in 1940 and, following the Japanese attack in 1941 on Pearl Harbour, in 1942 in America. Germany's motor industry had been cleverly rationalized by the state during the Thirties as a carefully orchestrated part of Germany's covert re-armament plans. Fewer companies were, however, diverted totally from vehicle production since there was a far greater need for trucks and general purpose vehicles, tanks, light gun carriers and half-track vehicles.

By 1939 the American automobile industry had been whittled down to

Designed in the late-Thirties but introduced in 1949, the diminutive Citroen 2CV was an ideal car for austere post-war days, and has been made ever since with few changes in fundamental concept.

just 18 makes of any significance, of which Chevrolet, Buick, Pontiac, Oldsmobile, and Cadillac-La Salle were controlled by General Motors; Ford controlled Lincoln-Mercury; Plymouth, Dodge, Chrysler and De Soto were all members of the Chrysler camp. This left only Willys-Overland, Studebaker, Packard, Hudson, Nash-Lafayette, Graham and Hupmobile independent, and the latter two were on their last legs.

In many respects, the war saved the American independents, and it certainly gave a new lease of life to Willys who, together with Ford, turned out hundreds of wartime Jeeps. This ultra-successful maid-of-all-work was, ironically, designed by the struggling Bantam company, originally formed to build the British Austin Seven under licence in America, but who had neither the facilities nor the capital to produce it in the numbers demanded by the military.

Such was the enormity of the 1939–45 conflict that it took civilian manufacturers quite a time to get back into stride. Even the Americans had only ageing 1942 designs to offer at first, with the exception of two, both significantly the very last to challenge the supremacy of Detroit mass production, for both Tucker and Kaiser were newcomers. The Tucker

A wartime maid-of-all-work, the Jeep, made by U.S. companies Ford and Willys. Now the GPW ¼-ton utility vehicle is a precious collector's item, after thousands were abandoned to scrap in the late-Forties. More than 635,000 Jeeps were made during the Second World War.

Torpedo, built in Chicago, Illinois was far ahead of its time, with padded dashboard, pop-out windscreen and front passenger crash compartment, disc brakes and flat-six, rear-mounted, water-cooled engine. After three stormy years, however, the project folded.

The Kaiser (and its sister car the Frazer) was built in the huge plant at Willow Run, Michigan, used for the mass-production of B-24 Liberator bombers. Remembered for their stylish and innovative designs, Kaiser lasted until 1954 when, having taken over Willys, they moved to Brazil as the Kaiser-Jeep Corporation.

The success story of the independents was surely Hudson and Nash, who together formed American Motors and showed the way towards the future in 1958 with the introduction of their first 'compact' car, considerably smaller than the huge, over-engined hippos which Detroit had been churning out since the war, and far more acceptable to European markets. It was a winner and, named the Rambler, sold

Successor to the Citroen *Traction Avant* of 1934, the equally advanced DS19 was first shown in 1955. Its front wheel drive and self-levelling suspension plus power-assistance for almost everything else caused a sensation.

Right: The motor industry's classic disaster? The Edsel, named after Henry Ford's son, was introduced in 1958 to fill a market gap that had vanished by the time it appeared. Production ceased late in 1959.

Below: Production of the Chrysler 'letter' range of cars started in 1955 with the 300 housing a V8 6½-litre unit. This is a 1960 300F a model with a number of NASCAR records to its credit.

486,000 units in 1960 to take American Motors into third place in the sales race, ahead of Chrysler's Plymouth.

American styling during the Fifties left much to be desired, and while engineering improvements saw the gradual disappearance of the straight-six in favour of the V-8 and increasing use of automatic transmissions, handling fell far behind European standards.

In Britain, post-war styling tended to be pure 1939 with bits tacked on. Separate wings and headlights and sit-up-and-beg bodywork were still the order of the day in 1946, together with petrol rationing and a quota

Right: Small Wonder. The VW Beetle changed little in appearance since it was first made in early post-war days. Its numbers outran the record 15 million Ford Model Ts and formed the basis for the sophisticated Porsche.

BGD-M 573

Successor to the XK120 and the competition Jaguars, the E-Type became a classic car overnight. This is a 1961 3·8-litre, one of the first off the line.

system which dictated that the majority of new cars went for export to earn much-needed foreign currency. Those who were lucky enough to be able not only to afford a new car but to obtain one before 1950, had then to sign a covenant that they would not re-sell it for two years.

Notable exceptions in the styling doldrums were, strangely, two firms who had previously been renowned for their conservative traditions. Armstrong-Siddeley had always occupied uneasily a market slot somewhere midway between Humber and the smaller Rolls models, and had advertised their pre-war 12 hp as being 'for the daughters of gentlemen', but by 1949 they boasted independent front suspension and were among the first to announce their post-war programme.

Jowett on the other hand, whilst remaining wed to their pre-war flat-four engine configuration, swept aside their agricultural image once and for all with a most attractive fast-back saloon, the Javelin, announced in 1947. At the time when everything else looked undeniably 1939-ish, it boasted unitary body-chassis construction, aerodynamic lines, all round independent suspension giving excellent handling, and a top speed of 80 mph. Garage mechanics complained that the front wheel had to be taken off to change a plug, but the class wins notched up by the make in the 1949 and 1951 Monte Carlo Rallies and at Le Mans in 1950, 1951 and 1952 silenced the critics.

Experience gained in motor sport (Jaguar won the Le Mans 24 Hours race no less than five times in the Fifties) has always been reflected in Jaguar production cars, a current model of which is this sleek 5343 cc V12 150-mph XJS.

The ubiquitous Mini. First seen in 1959 behind a Morris or an Austin badge, the tiny transverse-engined Mini soon gained a reputation of its own – mainly by winning almost any sport in which it competed.

THE POST-WAR WATERSHED

The year 1948 was in fact the watershed for post-war British design, heralding the introduction of the 'new line'. For manufacturers like Singer, Standard (with their Vanguard) and Austin this meant recessed headlights, integral wings [fenders] and a slabsidedness which owed much to Detroit and which did little to enhance appearances. It was, however, the first real step towards modernity, and while initially the new bodies often tended to cloak mechanics of pre-war origin (Singer's chain-driven overhead camshaft four dated from the late Twenties, and the new Morris Minor still utilized the pre-war side-valve eight) the trend was towards new engine designs as well as new styling.

Austin emerged as the dominant post-war giant. The merger of the two companies Austin and Morris to form the British Motor Corporation in 1952 was the logical outcome. Fitted with Austin's overhead valve A35 engine, the Morris Minor went from strength to strength and was not finally phased out until the Seventies.

Because Britain's bread-and-butter family cars were unsuitable for many overseas markets – particularly the United States – much of her exports to the U.S. in the early post-war years tended to be sports cars of the type for which Britain had become famous, and for which G.I.s stationed there during the war had acquired a taste. Of these, the 'T'

In 1976 Porsche offered a front-engined car, the 2-litre 924, for the first time in its history. The larger 4·5 928 appeared in 1978. This is the 924.

series MGs – the TD, TC and the TF – represented the logical development of the pre-war two-seater, and until 1956 they were largely responsible for the popularization of the British sports car in America.

But the old order was changing and in 1948 William Lyons, whose SS Jaguars had dominated the pre-war market for low priced yet stylish sports saloons and dropheads in Britain, introduced an entirely new sports model. Designated the XK 120 and boasting a twin cam 'six' of own manufacture, this revolutionary model incorporated all that was best in modern aerodynamics, and combined this with 160 bhp and a top speed in production form of 120 mph, at the very low price of £1,263.

An immediate success, it signalled the development of the post-war sports car. When MG went modern in 1955 with their MG 'A' they also offered a twin cam variant, but it was not successful and the husky 1·6-litre ohv unit proved just as fast. In 1962 the MG 'A' became the 'B' with integral body/chassis construction, by which time a baby MG named the Midget, after its pre-war counterpart and based on the Austin-Healey Sprite, was continuing the tradition of a small, cheap sports car in the family.

On the Continent, the pattern of development which had commenced

A Rolls-Royce Carmargue; eight-cylinder, 6·75-litre, with a top speed of 119 mph (190 km/h) and everything else man could desire.

in the Thirties dictated that the European car was considerably advanced in specification when compared to its post-war British and American contemporaries. Independent all-round suspension was commonplace, and two-stroke engines and front-wheel drive were combined by firms like DKW and Saab.

Citroën of France fielded the sublime and the slightly ridiculous side by side with their advanced DS19 aerodynamic saloons and dropheads on the one hand and with the pre-war designed 2CV 'corrugated iron garden shed' on the other. The strange shape of the first DS19 shocked the public when first announced in 1955, but it did not date, and its unique self-levelling air suspension and power-assisted brakes, steering and gear-change made it as ahead of its time as had been its predecessor, the 'Traction Avant', in 1934. The 2CV combined an uncompromising ugliness with a simplicity of design and operation and a frugality which immediately endeared it to the French. It is still with us today in very much the same form in which it was introduced.

The best that Ford of Britain could offer in the early Fifties was the archaic Eight and Ten (in Anglia and Prefect guise) and the V-8 Pilot – the nearest that Dagenham ever got to Detroit 'Bulboidism'. These were

UKJ 605
NOF 652R

Left: The Triumph TR2 was introduced in 1953, running through a range of numbers to today's TR7, with its lively 2-litre, 16-valve, four-cylinder unit.

Below: A compact with good performance, the U.S. Ford mustang appeared in 1964. It sold half a million in 18 months. This is a late-Sixties model.

swept away and replaced by the stylish Consul, Zephyr and Zodiac in 1956 – some of the best-looking Fords ever built – but it was not until 1963 that Ford brought out a real winner in the Cortina. Combining good looks and performance with reliability and acceptability to tuning, it was attractive to both family and sporting motorist. 'Tweaked' by Colin Chapman, and in twin cam form, it was capable of 100 mph plus as the Lotus-Cortina.

Not all of Ford's new models were successful, however, and the Detroit parent dropped in 1958 what was perhaps the most classic automotive brick of all time with the Edsel. Named after Henry Ford's eldest son, and the result of extensive market research, this unfortunately-styled disaster took months to get into production, by which time the market gap at

Above: A 1980 Pontiac Firebird Trans-Am. With power at around 6½ litres, it has high performance but is a heavy gas-guzzler.

An Aston Martin DB3S of 1955, still in mint condition and with just 15,000 miles on the clock! It develops 230 bhp from 2·9 litres.

which it was aimed had closed. Ford persevered with it for three seasons and then wrote off umpteen million dollars to experience.

Chrysler never really regained their pre-war status in America (they had been the fastest growing newcomer during the Twenties) and they pitched from high-hipped ugliness to extravagant fins and other panel-beaters' nightmares in the Fifties and Sixties. In an effort to stave off disaster they widened their horizons and expanded heavily into Europe, taking over Simca of France (who had themselves absorbed Ford of France), and the ailing Rootes Group in Britain (which encompassed Hillman, Humber, Singer, Sunbeam-Talbot, Commer and Karrier).

Not all cars which sold well when imported into America were of sporting or high performance, and one of the most successful European small cars to break into the U.S. was the Volkswagen. Conceived in the Thirties by Dr. Ferdinand Porsche as Hitler's 'Strength through joy' car,

First seen in 1953, the fibreglass-bodied Chevrolet Corvette quickly became the sporting man's automobile, the only production sports car in the U.S.A.

in basic Kubelwagen and Schwimmwagen form it was extensively used by the German Forces during the war.

This odd little vehicle was evaluated by Humber in the aftermath of the war in 1945–6. On paper, its specification was not impressive. A beetle-shaped body on a backbone frame was propelled by a noisy, rear-mounted, air-cooled, horizontally-opposed engine which, in full song, sounded like a bag of old nails. Humber, with predictable British phlegm, found that the design held 'no merit', and plans to build it in Britain were dropped. Production was eventually commenced in 1945 in Germany under British control, and 1,785 cars delivered to the British Army. The factory reverted to civilian control in 1949 and thus Britain lost out on the greatest success story of the post-war era.

Progressively updated and increased in size from the original 1,131 cc of the early model to a hefty 1600 cc, the Beetle swept Europe and

Above: Replica cars are attracting a larger market than ever before. A fine example is this Panther Westwinds J72, built in traditional style by traditional methods, and somewhat reminiscent of the SS100 on page 59.

Left: Race-bred and roadworthy, the Ferrari 308 GTB, with its modest 2·9-litre engine, is capable of 250 km/h (155 mph).

America, and generally won the affection of the motoring public everywhere. Not only did it remain in production until the late Seventies, achieving the rare distinction of being the only single model to outstrip the Model T Ford's 15,000,000 total production, it also formed the basis of the outstandingly successful Porsche sports cars. The 911 series was progressively developed in six-cylinder form and the 907 2·2-litre prototypes were 1st and 2nd in the 1968 Sebring 12-Hour Race, gained a 2nd at Le Mans and won both the Monte Carlo and the Swedish rallies.

The Lotus Eclat 2+2 is a logical development of the Lotus line, which has become more de-luxe though no less sportive during recent years. Engine: four-cylinder 2-litre, with 160 bhp at 6200 rpm.

THE FUEL-SAVING SEVENTIES

By the late Seventies, Volkswagen were pinning their hopes at the popular end of the market on the Golf and the Polo. The Polo, like the Ford Fiesta and three or four almost indistinguishable contemporaries, reflects the current preoccupation with fuel conservation and general economy, but owes its transverse-engined front-wheel-drive to a much earlier pioneer.

Like the earlier Morris Minor, the Morris Mini-Minor and Austin

Seven of 1959 were both designed by Alec Issogonis and soon, simply dubbed the Mini, revolutionized small car design. Fitted with a 848-cc transverse engine driving the front wheels, and with the gearbox living in the sump, this compact little car boasted hydrolastic suspension, exceptional cornering ability and handling, and a susceptibility to tune which made it an immediate favourite with sporting motorists. In Cooper 'S' form, with a 1275-cc engine, it could sometimes even see off Jaguars in competition.

Jaguar, quite apart from their sporting successes with the XK120 and its derivatives the XK140 and XK150, entered motor racing officially in 1950, winning Le Mans no less than five times, in 1951, 1953, 1955, 1956 and 1957. In the last of these events, the unitary construction 'D' types finished in incredible fashion in 1st, 2nd, 3rd, 4th and 6th places. The

experience gained in competition resulted in a very successful line of saloon cars. The Mk II was produced for 13 years and continued through the 420 and 'S' types, the 240, XJ6 and XJS to provide Jaguar's traditional formula of 'grace, space and pace'. It would certainly be correct to say that they never built a bad-looking car, and the 'E' type sports model caused a sensation when announced in 1961. With V-12 engine, the Series 3 'E' type took this ageless design well into the Seventies.

In the post-war period, take-overs were to be the order of the day in Britain. First to start the bidding were Standard, who purchased Triumph and developed their 1800 engine into the husky 2088-cc unit which powered both the Vanguard saloon and the solid and enduring little TR2 sports car. The David Brown tractor group absorbed body-

Below: This Aston Martin Lagonda represents the finest in British automotive electronic innovation and workmanship. Engine: 5·3-litre V8.

Below right: This 350SL Mercedes-Benz, renowned for its style and safety, was first seen in 1971. A sports-tourer two-seater convertible, its 3½-litre V8 unit gives it a top speed of 203 km/h (126 mph).

builders Tickford of Newport Pagnell, and Aston Martin and Lagonda, and transferred all operations there, giving W. O. Bentley's last design for Lagonda a new lease of life in both marques.

Leyland in turn took over Standard-Triumph, Jaguar bought Guy lorries and Daimler, Rover snapped up Alvis – keeping their military vehicle side going but phasing out their cars – and so it went on until, at last, the whole situation got completely out of hand. Then, under the direction of Donald Stokes – erstwhile super salesman of Leyland buses to China and Cuba – Leyland rapidly bought up everyone in sight and finished up with an enormously unwieldy empire which encompassed Rover, Riley, Austin, Morris, Wolseley, MG, Austin-Healey, Triumph, AEC, Albion, Guy, Jaguar, Daimler, et al.

Virtually the only companies to escape the take-over fever which pervaded the Sixties and Seventies were the very large, the very old, the very small and the exotic, and thus Rolls-Royce (despite the disastrous

The elegant Chrysler New Yorker 5th Avenue would grace any capital city. It is offered in four versions – a 2-door hardtop, a 2-door sedan, a 4-door sedan and a 4-door hardtop.

bankruptcy of the parent aero-engine company in 1971) have remained healthily and individually Rolls-Royce, still able to command in the 1980s the respect which earned them the title of 'The Best Car in the World'. Post-war production centred initially upon the Wraith and Mk6 Bentley on updated versions of the old straight-six engine. It was not until 1959 that this configuration gave way to a 6·2-litre V-8 on the Cloud II and the S2 Bentley. Highly sophisticated transmission had, however, been standardized on the Cloud I and this feature was continued with the introduction of the completely new Silver Shadow in 1965.

There have been few true American sports cars. Ford made the Thunderbird of the Fifties and the semi GT Mustang in the Sixties, but both lost their initial sporting image after a few seasons as they grew heavier non-sports bodies, while Chevrolet's Corvette, born in 1953, retained the sporting persona for over a decade. Both Pontiac and Oldsmobile – particularly the latter's Tornado – have fielded models which are agile rather than sporting, but with the growth in popularity of saloon-car racing this absence of pure sports models has not been so much felt.

Of the Continentals the sleek Lamborghini, like Aston Martin, owes its existence to tractor manufacture. The Italian company has fielded an incredible assortment of post-war models ranging from a four overhead camshaft V-12 developing 360 bhp, to the Marzal with a 2-litre, six-cylinder engine and six headlamps!

Ferarri have continued to command world-wide respect in the competition world in a way in which no other Italian car, past or present, can hope to emulate. From their first Le Mans win in 1949, through the World Championships they gained for drivers of the calibre of Fangio,

Hawthorn and Phil Hill, and the World Sports Car Championship wins of 1958, 1960 and 1961, they proved themselves again and again to be virtually invincible. While their Grand Prix fortunes have declined somewhat since the mid-Sixties, the early Seventies saw an amazing eleven consecutive wins in the Manufacturers' Sports Car Championship. Inevitably, Fiat now control the company, and the first fruit of the marriage – the Fiat Dino – boasted a Ferrari-designed engine.

Lancia, another old-established marque destined for Fiat control in 1969, have the enviable reputation of having never made a bad car. Their take-over was purely a matter of economics; in 1958 they had sold only 8794 cars to Fiat's 169,532. Happily, however, Lancia have retained for themselves a measure of independence, and the cars flying the Lancia flag represent the elite of the Fiat range. That this formula works was graphically demonstrated when, in December 1979, journalists voted the Lancia Delta – another transverse-engined small car – 'Car of the Year'.

The pooling of knowledge between rival companies and the production of components such as engines for common use is now increasingly popular. Both Peugeot and Volvo are involved in such schemes. To a lesser extent the theme is followed by the giants, with General Motors basing its small Chevrolets, Vauxhalls and Opels on a single master design, a distillation of technologies under one hood.

Oldest in the industry and still on top, Mercedes-Benz made a good recovery from the war years, and during the Fifties provided the car which won the Mille Miglia for Stirling Moss – the 8-cylinder, 3-litre 300 SLR – and indeed virtually every other major sports car event in 1955. The 300SL and its smaller cousin, the 190SL were both also highly successful as production models, and Mercedes have continued to

In the late Thirties American Bill Stout designed the bizarre Scarab, a vehicle with an interior like a living-room, in which the seats could be moved into various positions – even set up for card games. The streamlined body had no hood [bonnet] and a Ford V8 powerpack was housed at the back of the car. One of the few Scarabs actually built was used by General Eisenhower during his European campaign in the last war.

combine successfully sporting cars with high quality saloons and limousines – of which the 600 was the post-war equivalent of the enormous 'Grosser' beloved of Nazi leaders.

The years since the ending of the Second World War have been momentous and full of change. They have seen the ranks of the independents further decimated with only a few successful specialist newcomers, and the remaining makes brought together in massive groupings. The accent on safety has led most manufacturers away from open models so that now only the out-and-out sports car makers offer them. The world energy and fuel crisis, inflation, and the need for economy, has signalled the demise of the large and luxurious in favour of the efficient, compact, well-designed small car.

Now with two spectacular factors of the Seventies and Eighties – the risen sun of Japan, today in a strong position in the world's automobile markets – and the increasingly common automation of all facets of car-producing (such as Fiat's Strada which is produced almost untouched by human hand), the direction of car manufacturing is likely to be even more a matter for computer operators and promotion campaigns.

New techniques will engender new standards, and new restrictions on the use of fuel will call for new philosophies. Our new concepts of what constitutes adequate safety will also (perhaps even primarily) influence our conception of what constitutes a classic car. Certainly the accent is likely to be more on arriving at your destination in comparative safety than in the doubtful joys of travelling.

Just like it was when that first reliability trial was set up by *Le Petit Journal* in 1894.

Englands Ende in Der Schlacht Bei Dorking: Erinnerungen Eines Alten Britten Im Nächsten Jahrhundert

George Tomkyns Chesney, Wilhelm Hübbe-Schleiden

ENGLANDS ENDE

in der

Schlacht bei Dorking.

Erinnerungen eines alten Britten im nächsten Jahrhundert.

Eine Studie,

aus dem Englischen übertragen

vom Verfasser des

ETHIOPIEN.

Mit einer Karte.

HAMBURG.

KARL GRÄDENER.

1879.

The author of the original (imaginary) Battle of Dorking General George Chesney, (M.P. for the city of Oxford) died in London 1895 -

Wir Engländer sollten wachen und schaffen. — Ermüden wir jetzt und träumen selbstzufrieden dahin, so werden wir eines Tages unsanft erweckt werden mit der Thatsache, dass unsere Zeit dahin ist, und dass in dem Wettstreit europäischer Grossmächte Deutschland die erste ist und der Rest — nirgends!

(*Macm. Mag.* July 1878, 221.)

Ihr Enkel wollt, dass ich euch erzählen soll, wie es denn zugegangen ist, dass vor nun bald funfzig Jahren dieses Elend über uns hereinbrach. — Nun, ich habe freilich selbst mein hartes Theil mit davon erlebt; aber wahrlich, keine Arbeit ward mir noch so sauer, als zurückzudenken an die traurigen Zeiten — an das trostlose Ende von Alt-Englands Weltmacht.

Doch es mag auch gut sein, dass ihr's hört und leset; und wenn ihr jetzt von hier fortzieht in die Ferne, mag es euch eine Lehre sein, da wo ihr euch eure neue Heimath gründen möget. — Für uns kam diese Lehre nur zu spät; und doch waren auch wir genug gewarnt, wenn wir nur hätten sehen und hören wollen. Die Gefahr kam uns nicht einmal eigentlich unerwartet: Sie brach plötzlich über uns herein, das ist wohl wahr; doch ihre Vorboten hatten lange deutlich uns vor Augen gestanden; wir aber waren blind mit offnen Augen. Niemandem anders als uns selbst haben wir diese Schmach zuzuschreiben, in der Gross-Britannien zu Grund ging.

O, diese Alte Gute Zeit! — Die Alte Böse Zeit, sollte ich sagen, die ein Mannesalter wie das unsere entehrt sah. Ja, selbst jetzt noch, wo doch funfzig Jahre fast darüber hingegangen sind, kann ich nicht ruhig einem jungen Manne iu die Augen sehn, wenn ich zurückdenke, dass ich Einer war von denen, deren Jugend für ewige Zeit gebrandmarkt ist mit dem Sturz Alt-Englands — Einer von denen, welche diesen grossen Namen schändeten, der uns von unsern Vätern unbefleckt überkommen war.

Wie glücklich war doch dieses Land damals! — Mit dem System unseres Freihandels hatten wir den Verkehr der ganzen Welt seit fast einem halben Jahrhundert beherrscht, und der Reichthum, der uns zufloss, schien kein Ende nehmen zu wollen.

London wuchs und wuchs; kaum konnte man schnell genug Häuser bauen für all die reichen Leute, die hier wohnen wollten, für die Kaufleute, die in den verschiedenen Theilen der Welt Geld verdient hatten und dann zurückkehrten, um sich hier ihr Heim zu gründen, und die Staatsbeamten, die Gelehrten, Ingenieure und wer sonst noch! Auch die Krämer und die Arbeiter hier hatten damals Theil an dem allgemeinen Wohlstand. Die Strassen reichten bis hinaus nach Croydon und nach Wimbledon, die zu meines Vaters Zeiten nur noch kleine Landstädte waren; und man sprach schon von der Zeit, dass Kingstone Reigate und Dorking auch bald mit zur Stadt gezogen werden würden. Wir dachten damals, wir könnten nur so fortfahren zu bauen und uns auszubreiten, Reichthümer zu sammeln, und ganz unbeschränkt damit zu wirthschaften. Freilich fehlte es auch damals nicht an Armuth; die Leute, die kein Geld hatten, vermehrten sich gerade so schnell wie die reichen, und der Pauperismus fing schon an uns zu bedrohen; aber wenn auch damals wohl die Armensteuer hoch stieg, so war doch immer reichlich Geld da, sie zu zahlen. Und nun gar der Mittelstand! Es schien wirklich, als ob seine Zunahme und seine Prosperität keine Grenzen haben könne. Ein Dutzend Kinder in die Welt zu setzen, hielten diese Leute damals für etwas ganz Natürliches — oder wie man sich damals ausdrückte — *die Vorsehung sandte ihnen diesen Kindersegen;* und konnte man auch nicht immer alle Töchter unter die Haube bringen, so fand man doch stets wenigstens ein gutes Unterkommen für die Söhne. Da war genug zu thun in allen Berufsarten; auch waren damals sehr leicht Staatsanstellungen zu haben in den vielen Zweigen unsrer staatlichen und städtischen Behörden, die sich mehr und mehr ausdehnten, und an Macht und Einfluss zunahmen. Ueberdies konnten wir damals junge Leute noch nach Indien schicken, wo sie was Rechtes werden konnten, oder sie gingen zum Militär oder zur Marine. Damals zogen sogar ganze Familien frohen Muthes hinaus nach irgend einer jener reichen Colonien, die wir in allen Theilen der Welt hatten, und wo sie wussten, dass sie Freunde und Verwandte dort in Glück und Wohlstand finden würden, und wo sie sich dann auf dem fernen brittischen Boden eine neue Heimath gründen konnten — nicht so wie jetzt, wo wir auswandern, weil wir an Alt-England verzweifeln und nun heimathslos hinausziehn müssen in die weite,

weite Welt, die jetzt doch nur dem andern Volk gehört. Selbst Schullehrer machten damals gute Geschäfte hier zu Lande. Freilich lehrten sie n cht gerade viel — gewiss nicht, denn sie wussten selbst nicht übermässig viel, und sie brauchten damals auch nicht annähernd so vielerlei zu wissen, wie man jetzt unter dieser Regierung von ihnen fordert; aber sie verhungerten doch damals trotz ihres blossen *gesunden Menschenverstandes* nicht, denn überall im Lande baute man grosse neue Schulen mit vier- oder gar fünfhundert Schülern.

O, wir Thoren, die wir waren! Wir dachten all der Wohlstand und das Glück sei uns von Gott bestimmt und könne nimmer enden. Wir waren so blind nicht einzusehn, dass unser Land nur eine grosse Werkstatt war, in der wir die Erzeugnisse fremder Länder und Erdtheile verarbeiteten, und dass wenn unser Weltverkehr einmal gewaltsam unterbrochen würde, und ein andres Volk in der Zwischenzeit diese selben Dinge billiger und besser fabriciren würde, dann auch Niemand weiter unsre Fabricate kaufen und uns das viele Rohmaterial schicken würde, sondern nur — wie jetzt — dem andern Volke, das uns überwunden hat; und hier in unserm Lande selbst Baumwolle und dergl. bauen, können wir doch auch nicht. Es täuschte uns vornehmlich damals jener grosse Vortheil, den wir durch die Menge und die Billigkeit unsres Eisens und unsrer Steinkohlen hatten: Aber schon damals machte es sich nns fühlbar, dass in andern Ländern viel mehr Kohlen sind und bessres Eisen; und schon damals waren beide Artikel anderwärts billiger zu haben als bei uns.

Aber Dank des grossen Weltgetriebes, das wir nun einmal in der Hand hatten, dachten wir nicht daran Haus zu halten; unbesorgt um die Zukunft sandten wir in alle Welt hinaus, was wir jetzt von andern Ländern kaufen müssen.

Und in Betreff der Nahrungsmittel war England damals auch nicht reicher als wir jetzt sind; aber darum zehrten wir doch damals nicht an so kärglichem Bedarf wie jetzt. Im Gegentheil unsre weniger bevölkerten Nachbarländer lieferten uns stets das Beste und das Schönste, was sie hatten. Lediglich deshalb nur waren wir damals so reich, weil die anderen Nationen in allen Theilen der Welt durch das grosse Wirthschaftsnetz, welches wir über die ganze Erde gespannt hatten, veranlasst waren, ihre Producte nur zu uns zu senden, um sie zu verkaufen, oder sie bearbeiten zu lassen. Und wir Thoren dachten, das würde immer so bleiben. — — Vielleicht hätte es so bleiben können, wenn wir zum Mindesten nur die Anstalten danach getroffen hätten. Aber wir vergassen, dass auf unsre Macht allein der Weltverkehr sich stützte, der die Quelle unsres Reichthums war, und wer von uns dies auch wohl fühlen mochte, dachte doch nicht anders als Britanniens Weltmacht sei ganz unbesieglich. In unserm Leichtsinne waren wir zu sorglos, um daran zu denken, unsre günstigen Verhältnisse zu sichern und zu schützen als wir es noch konnten. Als aber erst einmal der Weltverkehr uns entzogen war, wollte es uns ohne unsre frühere Macht nicht mehr gelingen, ihn uns wieder zu gewinnen.

Und doch, wenn je ein Volk gewarnt war, so waren wir es. — Wie wir einst die grösste Handelsmacht der Welt waren, so waren unsre Nachbarn jenseits des Canals einst die stärkste Militärmacht in Europa. Auch bei ihnen blühte der Handel, und sie galten gar in mancher Hinsicht als das erste Volk Europas; aber auf ihre Armee bildeten sie sich doch am allermeisten ein. Und mit Recht! Sie hatten einst die Russen und die Oestreicher geschlagen, und vor Zeiten auch sogar die Deutschen, darum meinten sie, sie wären ganz unüberwindlich. — Ich entsinne mich noch sehr wohl der grossen *Revue*, die Napoleon III. in Paris abhielt während einer grossen Ausstellung (es war 1867) und wie hochmüthig er sich da gebehrdete in dem Bewusstsein, all den anwesenden Königen und Fürsten seine famose Garde vorzuparadiren. Und, drei Jahre später — wurde diese selbe Macht, die solange als die erste in Europa galt, elendiglich geschlagen und fast alle diese prachtvollen Soldaten zu Gefangenen gemacht. Solche Niederlage war bis dahin noch ganz unerhört in der Weltgeschichte; und mit solchem Beispiel vor unsern Augen, sollte man glauben, hätten wir soviel Sinn und Verstand haben müssen, diese Lehre zu beherzigen. Aber nein, wir glaubten nicht an die Möglichkeit eines solchen Unglücks — lediglich deshalb, weil uns selbst nie dergleichen betroffen hatte.

Freilich rührte es sich auch in unserm Volke damals eine Weile, und man meinte unsre Heeresmacht müsste wohl reorganisirt werden, man solle auf Vertheidigungsmittel sinnen, und sich sichern gegen solche Schlagfertigkeit der gewaltigen Macht, die solche unversehnen Angriffe mit dieser Leichtigkeit bewerkstelligte. Unsere Regierung brachte damals sogar den Entwurf einer Militär-Reform beim Parlamente ein. Es war das, wenig gesagt, eine halbe Massregel, und unglücklicher Weise wurde überdies der Entwurf nicht vom Parlamente als eine nationale Angelegenheit aufgenommen, sondern zur Parteisache gemacht — und fiel durch. Es war eine Radicale Partei im *House of Commons,* deren Stimmen nur per Compromiss zu sichern waren, und die blindlings als den Preis ihrer Zustimmung gerade eine Reduction unserer Heeresmacht, wo nicht gar vollständige Abrüstung forderte. Es war die stationäre Politik dieser Partei, alles militärische Wesen in Verruf zu bringen, um dadurch, wie sie meinten, den Einfluss der Regierung und der Aristokratie zu brechen. Diese Leute wollten nicht begreifen, dass die Zeiten sich schon damals längst geändert hatten, dass die Regierung bei uns thatsächlich gar keine Macht mehr hatte, und in Wirklichkeit ganz dem Belieben des *House of Commons* Preis gegeben war, und dass sogar diese Parlaments-Macht anfing, einer Pöbelherrschaft Platz zu machen. Schliesslich wurden die Minister von allen Seiten so in die Enge getrieben, dass sie nach und nach alle wesentlichen Punkte des Entwurfs fallen lassen mussten — und mit denselben war es ihnen auch kaum Ernst gewesen. Es fehlte damals übrigens keineswegs an Geld, das zu Militärzwecken disponibel war, wenn es nur in der rechten Weise aufgewendet worden wäre. Die Armee kostete uns wahrlich genug, ja mehr als genug, um uns eine völlig ausreichende Vertheidigung zu ermöglichen, und wir hatten Leute ausgerüstet mit Waffen aller Gattungen — mehr als wir brauchten, wenn sie nur verständig organisirt gewesen wären. Gerade in der Organisation und in Vorsichtsmassregeln liessen wir es aber am meisten fehlen, und die massgebenden Leute bei uns glaubten nicht an die Nothwendigkeit solcher Vorkehrungen.

Unsere Flotte und der Canal, der uns vom Continente trennt, meinte man, sei genug Schutz für uns. So wurde diese *Militär-Reform* denn aufgeschoben bis zu *mehr gelegner* Zeit, und die *Miliz* und unsere *Voluntiers* wurden nach wie vor ganz sich selbst überlassen; *sie einexerciren zu wollen, wäre,* so sagte man, *eine Einmischung in das Recht der Selbstbestimmung unsrer Bürger gewesen.* — Aber was soll ich euch mehr von diesen Zuständen erzählen, die euch so oft schon dargestellt sind?

Obwohl die Nation sich unbehaglich genug fühlte, so wurde sie doch erfolgreich getäuscht durch das angebliche Gefühl der Sicherheit, welche die tonangebenden Stimmen zur Schau trugen; die Warnung, welche uns das Unglück Frankreichs gab, ging an uns unbeachtet vorüber. Man hielt es damals bald sogar nicht einmal mehr für nöthig, unsre Arsenale an sichere Plätze zu verlegen, oder doch wenigstens unsre Hauptstadt gegen eine Ueberrumpelung zu sichern, obwohl die Kosten solcher Anstalten nicht einmal höhere Steuern nöthig gemacht haben würden. Die Franzosen bauten auf ihr Heer und dessen Weltruf, wir auf unsre Flotte; und in beiden Fällen war das Resultat solches blinden Vertrauens ein Verderben, wie es unsre Väter selbst in ihren schwersten Zeiten sich nicht hätten vorstellen können.

Ich brauche euch hier kaum auseinanderzusetzen, was den schliesslichen Zusammenbruch veranlasste. Die Kämpfe in Indien nahmen einen beträchtlichen Theil unsres kleinen Heeres in Anspruch; dann kamen die Verwicklungen mit Amerika, die uns schon seit Jahren bedroht hatten, und wir sandten 10,000 Mann, um Canada zu schützen — eine Handvoll, die dort nicht einmal irgend etwas thun konnte zu einer wirklichen Vertheidigung des Landes, aber doch den Amerikanern zur unwiderstehlichen Versuchung wurde, auszuziehen, und sie alle gefangen zu nehmen, besonders da drei Bataillone unsrer stolzen Garde darunter waren. Auf diese Weise war die reguläre Armee hier im Lande sogar noch bedeutend kleiner als gewöhnlich, und von diesem kleinen Reste wurde wieder die Hälfte nach Irland geschickt, um eine Fenier-Invasion zu unterdrücken, die angeblich im Westen ausgerüstet sein oder werden sollte. Schlimmer noch als alles dieses aber war — doch weiss ich nicht, ob selbst dies von viel Belang gewesen wäre, wie die Sachen schliesslich sich gestalteten — unsre Flotte war über die ganze Welt zerstreut: Einige Schiffe waren in West-Indien zur Vertheidigung unsrer dortigen Colonien, andre waren in den chinesischen Meeren, um der amerikanischen Kaperei unsrer Schiffe dort zu steuern, und ein grosser Theil war im Norden des stillen Oceans, um die Küsten unseres Gebietes in Amerika zu schützen. Dies war damals schon ein sichtlich thörigter Versuch, und doch war Amerika vor 40 Jahren noch nicht die grosse Macht, welche es heute ist. Aber wir hätten uns schon damals sagen können, dass es ganz nutzlos sei, den Amerikanern diese Gebiete streitig zu machen, die doch durch die geographische Lage, und die Umstände der Zeit am natürlichsten von ihnen verwaltet werden konnten.

Um diese Zeit war's, als gerade unsre Schiffe über die ganze Welt zerstreut und unser kleines Heer in lauter kleinen Detachments in den verschiedenen Ländern der Erde zertheilt waren: da wurde der Geheime Vertrag publicirt, durch den Dänemark und Holland annectirt wurden. Man sagt jetzt wohl, dass wir aller Noth hätten entgehen können, wenn wir damals nur uns ruhig gehalten hätten, wenigstens bis alle andren Gefahren vorüber gewesen wären; aber wir Engländer waren immer unverbesserliche Trotzköpfe: Das ganze Land schäumte vor Wuth über diesen deutschen Handstreich; das Gebräu unsrer Tagespolitik in allen Zeitungen kochte über vor Entrüstung; und das Government, aufgehetzt durch den Pöbel, liess sich vom Strome treiben und erklärte Krieg. — Wir hatten es bis dahin noch immer fertig gebracht, uns durch durch alle Verlegenheiten hindurch zu helfen, und so zweifelten wir nicht, dass unser guter Stern und unsre Dreistigkeit uns auch dieses Mal schon durchhelfen würden.

Ihr könnt euch kaum einen Begriff machen von dem Schreien und Lärmen, das damals über das ganze Land losbrach. Nicht etwa, dass die Einberufung unsrer Reserven solche Unruhe verursacht hätte, denn ich glaube, da waren kaum 5000 Mann, die man hätte einberufen können, und ein grosser Theil von diesen konnte nicht einmal aufgefunden werden; aber dagegen rekrutirte man stark überall im Lande, und man zahlte ganz fabelhaft hohes Handgeld, denn bis

50,000 Mann waren jetzt endlich vom Parlamente bewilligt worden. Dann wurde ein *Aushebungs-Gesetz* erlassen, welches die Miliz um 55,500 Mann vermehrte; warum keine runde Zahl festgesetzt wurde, weiss ich nicht zu sagen, aber der Premier-Minister behauptete, dies sei die genaue Anzahl Leute, welche zur Vertheidigung des Landes genügen würden. Dann aber hättet ihr erst die Schiffbauerei sehen sollen, die nun losging! Panzerschiffe, Courierdampfer, Kanonenböte, Monitors — wurden massenhaft in Arbeit genommen. Jede kleine Werfte sogar erhielt ihren Theil an den Aufträgen hierzu, und man bot Tagelohn bis zu 10 ℳ. für jeden Arbeiter, der einen eisernen Bolzen einzuschlagen verstand. Dies war der Rekrutirung freilich nicht gerade förderlich, wie ihr wohl denken könnt. Ich erinnere mich auch, da war viel Gerede im Parlamente ob Handwerker zur Miliz gezogen werden dürften, da sie doch so sehr für den Schiffbau nöthig seien, und ich glaube, man nahm sie auch schliesslich nicht. Alles concentrirte sich auf die Werften; und ich zweifle nicht, wenn wir ein paar Jahre Zeit gehabt hätten statt der zwei, drei Wochen, die wir hatten, so würden wir wohl alles ganz vortrefflich fertig gebracht haben.

Es war ein Montag, als der Krieg erklärt wurde, und in ein paar Stunden hatten wir bereits den ersten Beweis davon, was für Anstalten von unserm Gegner für diesen Kriegsfall vorbereitet waren, den er in der That verursacht hatte, obwohl die eigentliche Erklärung von uns ausging: Ein frommer Aufblick zum *Allmächtigen Lenker aller Schlachten, dessen höchsten und allweisen Rathschluss wir herausgefordert hätten,* wurde uns zurück-telegraphirt; und von dem Augenblicke an war all unsere Verbindung mit dem Norden von Europa abgeschnitten. Unsre Botschafter und Gesandten wurden an demselben Tage peremptorisch auf und davon geschickt, und wir waren plötzlich ganz wie in das Mittelalter zurückversetzt.

Denkt euch nur das stumme Erstaunen, welches am andern Morgen in ganz London herrschte, als die Zeitungen ohne alle Tagesneuigkeiten erschienen; erschreckt, über das Ungeheure, was geschehen, wagten sie kaum den wahren Sachverhalt in Vermuthungen anzudeuten. — Das Ganze war in der That ein Krieg von Ueberraschungen für uns. Alles war von unsern Feinden vorher wohl überlegt und arrangirt; aber wir hatten eigentlich keinen Grund überrascht zu sein, da wir doch diese selbe Gross-Macht ein paar Jahrzehnte früher in wenigen Tagen hatten eine halbe Million Menschen in's Feld stellen sehen, mit denen sie die grösste Militär-Macht Europas ohne Umstände über den Haufen warf und davon nicht mehr Aufhebens machte als unser Kriegs-Ministerium für den Transport einer Brigade von Aldershot nach Brighton zu machen pflegte; — und überdies waren die Franzosen ohne all die Bundesgenossen besiegt worden, welche sich diesem Kriegszug gegen uns gezwungen anschlossen. — Das, was wir nun erlebten, war durchaus nicht ungewöhnlicher als das, was den Franzosen geschehen war, aber die Leute hier zu Lande konnten sich nicht an den Gedanken gewöhnen, dass was Alt-England nie vorher passirt war, nun mit einem

Male geschehen könne. Gerade wie unsre Nachbarn, wurden auch wir erst klug, als es zu spät war.

Natürlich dauerte es doch nicht lange, dass unsre Zeitungen wieder *Neuste Nachrichten vom Auslande* brachten — selbst die allmächtige Organisation, welche damals auf dem Continent im Gange war, konnte unsere geheimen Kriegs-Correspondenten dort nicht ganz ausschliessen; und obwohl quer durch Europa alle Telegraphen und Eisenbahnen gesperrt waren, so sickerten doch die Haupt-Thatsachen durch. Die gesammte Schifffahrt von Königsberg bis nach Bordeaux war mit Beschlag belegt. Die vereinigten Flotten der beiden grossen Mächte waren ausgerückt, und man sagte, sie hätten sich vor dem grössten Hafen des Nordens an der Elbe gesammelt; Truppen wurden an Bord aller Dampfer geschafft, die man in den continentalen Häfen fand; dies waren überdies meist brittische Schiffe. Es war klar, dass man eine Invasion beabsichtigte. Selbst damals aber hätten wir uns noch retten können, wenn nur unsre Flotte zur Hand gewesen wäre. Die feindlichen Batterien der Häfen, welche die zusammengestoppelte Flotte am Continent beschützten, wären freilich wohl zu stark für uns gewesen, aber nur ein oder zwei unsrer Panzerschiffe so gehandhabt, wie brittische Seeleute sie damals zu handhaben verstanden, hätten diese feindlichen Transportschiffe sämmtlich in Grund und Boden schiessen oder doch wenigstens etwas aufhalten können, dass wir so viel Zeit gewonnen hätten, wie wir brauchten um uns zu besinnen. Aber auch von unsrer europäischen Flotte war der stärkste Theil an den Dardanellen stationirt und anderweitig im Mittel-meer vertheilt, und ein paar kleinere Schiffe nnsrer Canal-Flotte, die noch übrig waren, beschäftigten sich an der Westküste von Irland eifrig damit, die vermeintlichen Fenier-Banden ausfindig zu machen. So gingen zehn Tage drüber hin, ehe wir überhaupt eine Art von Flotte zusammenbrachten; da aber war es bereits evident, dass die Anstalten unsrer Feinde viel zu weit vorgerückt waren, als dass wir sie noch durch einen Handstreich hätten vereiteln können. Unsre Nachrichten, die fast alle über Spanien kamen, waren meist sehr alt, ehe wir sie erhielten, und waren überdies nur unbestimmt und allgemein gehalten. Soviel aber wussten wir, dass wenigstens ein paar Mal hundert tausend Mann auf die Beine gebracht waren; zum Theil waren sie schon eingeschifft, und in wenigen Tagen sollten alle an Bord jener Transport-Dampfer sein; auch hatten unsere Feinde mehr Panzerschiffe zur Beschützung dieser Dampfer, als wir damals zur Verfügung hatten. Ich glaube es war die Ungewissheit über den Platz, welchen der Feind zu seiner Landung wählen würde, und die Furcht vor der Möglichkeit die feindlichen Schiffe zu verfehlen, welche unsre Flotte noch einige Tage an unsrer Küste festhielt; erst am Dienstag in der zweiten Woche nach unsrer Kriegserklärung, lichtete sie die Anker und dampfte in die Nordsee hinaus.

Ihr habt ja oft gehört, wie unsre damalige Königin am Tage vor dem Auslaufen die Flotte bei Dover besichtigte, wie sie in ihrer Yacht um die prachtvollen Schiffe herumfuhr und an Bord des Admiralschiffs ging, um von dem commandirenden Admiral Abschied zu nehmen; wie sie von Rührung überwältigt zu ihm sagte, dass sie beruhigt sei in dem Gedanken, dass die Sicherheit des Landes seinen Händen anvertraut sei. Ihr wisst ja, was für eine stolze Antwort ihr der brave alte Seemann gab, und wie die sämmtlichen Mannschaften, die auf den Raaen der Schiffe aufgestellt waren, in endlosen Jubel ausbrachen, als Ihre Majestät davonfuhr. — Diese Scene ward sofort nach London telegraphirt, und die Begeisterung der Flotte theilte sich der ganzen Stadt mit. Ich stand vor dem Charing-Cross Bahnhof, als der Extra-Zug der Königin von Dover dort ankam. Als Ihre Majestät in ihrer Equipage die Station verliess, war da ein Jubeln und Jauchzen, dass man hätte glauben sollen, wir hätten mindestens schon einen grossen Sieg errungen, wenn nicht gar den ganzen Krieg schon beendet. Das radicale Blatt, welches von jeher stark für Abrüstung agitirt hatte, war während der letzten 14 Tage sehr kleinlaut geworden und hatte schon verschiedene Möglichkeiten eines Compromisses empfohlen, um nur den Krieg abzuwenden. In seiner Ausgabe vom nächsten Morgen aber, war auch dieses Blatt schon wieder oben auf. Da hiess es unter anderm in einem wunderschönen Leit-Artikel: *Elende Zweifler fragen noch, wie wir der Invasion begegnen sollen? — Wir antworten, dass eine Invasion ganz unmöglich ist. Eine brittische Flotte, bemannt mit echten Britten, deren Muth und Enthusiasmus Rückhalt findet im gesammten Volke dieses Landes, ist schon unterwegs den frechen Feind zu demüthigen. Der Ausgang eines Kampfes zwischen brittischen Schiffen und denen irgend einer andern Nation, unter auch nur annähernd gleichen Chancen, kann nie zweifelhaft sein. England erwartet mit vollkommner Ruhe den Erfolg der bevorstehenden Schlacht.* — Das waren die Worte dieser vielgelesenen Zeitung, aber so dachten wir damals eigentlich Alle.

Am Dienstag den 10. August stach unsre Flotte in See. Sie nahm einen unterseeischen Telegraphen mit sich, den sie hinter sich auslegte, während sie vorrückte, so dass sie ununterbrochen mit London in Verbindung blieb, und die Zeitungen brachten alle paar Minuten Extra-Blätter mit den *Neusten Nachrichten.* Es war dies das erste Mal, dass man ein solches Kunststück ausführte, und man betrachtete es damals als ein gutes Omen, dass dasselbe glückte. Ob es wahr ist, dass unsere Admiralität in London diesen Kabel-Telegraphen benutzte, um vom Lande aus ganz widersinnige Contre-Orders nach der Flotte hinzutelegraphiren, kann ich nicht mit Bestimmtheit sagen; aber es wurde damals behauptet, dass auf diese Weise dem Admiral das Commando thatsächlich ganz unmöglich gemacht worden sei; und alles, was uns dann noch von der Flotte telegraphirt wurde, waren nur ein paar kurze Anzeigen, aus denen zuletzt weder die Admiralität noch sonst irgend Jemand klug werden konnte: *Dies Schiff ist auf Recognoscirung ausgeschickt; jenes Schiff ist wieder zur Flotte gestossen; die Flotte befindet sich jetzt auf Länge und Breite so und so;* u. s. w.

Dies ging so fort bis zum Donnerstag Morgen. Ich war gerade mit meinem gewöhnlichen Morgenzuge von Surbiton zur Stadt gekommen und war auf dem Wege nach meinem Bureau, da höre ich plötzlich die Zeitungsjungen ausschreien: *Neustes Extra-Blatt*

— *Der Feind ist in Sicht!* — Nun könnt ihr euch die Aufregung in London denken! Das Geschäft in Lombard-Street ging freilich seinen Gang weiter, denn die Wechsel wurden trotz aller Kriegserklärungen fällig, und obwohl über die Existenz des Landes gewissermassen vor unsern Augen jetzt entschieden werden sollte, so gab es doch immer noch Speculanten, die sogar hieraus ein Geschäft für ihren Geldbeutel machten. Aber zuletzt überwog doch das Interesse für die Flotte auch bei solchen Alltags-Menschen, die sonst nur für ihr eigenes Vermögen Sinn hatten. Sogar die Leute, welche nach den Bankcassen gingen, um ihr Geld einzuzahlen oder Cheques einzulösen, hielten sich damit auf, mit den Bank-Commis über die Neusten Berichte vom Kriegsschauplatze zu sprechen. In den Strassen konnte man kaum vorwärts kommen, so gross waren die Menschenmassen, die sich überall anhäuften, um die Extra-Blätter zu kaufen und zu lesen. In den Häusern suchten die Bureau-Chefs und die Prinzipäle sich unnöthig in den gemeinsamen Arbeits-Räumen ihrer Angestellten zu thun zu machen, als ob sie sich scheuten, einsam in ihren Privat-Zimmern zu sitzen, und alle paar Minuten wurde Jemand hinausgeschickt, um die neusten Berichte von der Flotte zu holen. So wenigstens sah es in dem Verwaltungs-Bureau aus, in welchem ich angestellt war. Still zu sitzen war ebenso unmöglich, wie irgend eine ordentliche Arbeit vorzunehmen, und wer sich nur irgend losmachen konnte, ging auf die Strasse und irrte unter der Menge umher in dem Gedanken, dadurch eher Neue Nachrichten zu erfahren. — So schlimm auch die Zeiten waren, welche nachfolgten, so meine ich, war doch die unerträgliche Spannung, in der wir an jenem Tage gehalten wurden und der Schlag, der darauf folgte, fast das Schlimmste, was wir in der ganzen Zeit durchzumachen hatten.

Ungefähr um 10 Uhr war das erste Telegramm gekommen; eine Stunde später kam eine neue Depesche, *dass der Admiral eine Schlachtlinie formiren lasse,* und bald darauf, *dass er Befehl gegeben habe, auf den Feind zu halten und ihn anzugreifen.* — Um 12 Uhr kam die Anzeige, *Flotte eröffnet Feuer ca. 3 Meilen unterm Winde von uns* (nämlich von dem Schiffe, das den Kabel legte). So weit war alles nur Erwartung, dann aber kam die erste Unglücksbotschaft: *Ein Panzerschiff ist in die Luft geflogen — die Torpedos der Feinde richten grossen Schaden an — die Admiral-Fregatte scheint zu sinken — der Vice-Admiral hat signalisirt, es soll* — dann schwieg der Kabel; und, wie ihr wisst, hörten wir weiter nichts bis am Tage darauf das einzige Panzerschiff, welches der furchtbaren Zerstörung entgangen war, in den Hafen von Portsmouth einlief. Dann erst erfuhren wir, was geschehen war: — Wie unsre braven Seeleute versucht hatten, die feindlichen Schiffe zu entern, und wie diese, sobald unsre Flotte sich ihnen genähert hatte, plötzlich vor dem Winde davongedampft waren, als ob sie die Flucht ergriffen, in Wirklichkeit aber nur um eine Unzahl von Torpedos hinter sich zurückzulassen, die eines unsrer Schiffe nach dem andern in die Luft sprengten. Dies alles war in wenigen Augenblicken geschehen.

Unsere Regierung schien schon vorher Nachricht von dem allen erhalten zu haben; das Publicum aber blieb mittlerweile in der unerträglichsten Spannung. Vergebens sehnte jeder sich, Gewissheit zu erhalten; es war an jenem Donnerstage ganz unmöglich, irgend einer Beschäftigung nachzugehen. — Ich war früh nach Hause gegangen zum gewöhnlichen Regiments-Exerciren. Diese Uebungen waren eine wahre Erholung in solcher peinlich unruhigen Stimmung; als sie aber vorüber waren, war ich wieder rathlos, was ich mit mir anfangen sollte. Ich fuhr wieder zur Stadt, irrte dort, wie die meisten andern Leute, rastlos umher und blieb zuletzt auf meinem Bureau; aber die Nachrichten von der Flotte, auf die ich wartete, kamen nicht. In dieser aufgeregten Unentschlossenheit verfehlte ich dann auch den Mitternachtszug vom Waterloo-Bahnhofe, und ging schliesslich zu Fuss nach Surbiton hinaus. — Es war eine heisse, schwüle Sommernacht. Kurz vor Sonnenaufgang langte ich endlich bei unserm Hause an. Die Strassen der Vorstadt lagen noch wie im Morgenschlummer; — es war die letzte Stille vor dem hereinbrechenden Sturme. Ich hatte meinen Schlüssel zu der Hausthüre und brauchte Niemanden zu stören. Leise ging ich die Treppe hinauf nach meinem Zimmer. Die Hausgenossen schienen ruhig zu schlafen, und kein Laut unterbrach das Schweigen dieser Morgenruhe, nur leise hörte man die Vögel im Garten singen und zwitschern. Mich ergriff der Contrast dieses stillen Friedens mit der leidenschaftlichen Erregung, die mit diesem neuen Tage wieder losbrechen musste. Das Haus lag ganz so still und friedlich da, wie ich es oft in jener glücklichen vergangenen Zeit gefunden hatte, wenn ich von Gesellschaften oder Bällen spät nach Hause kam. Obwohl ich herzlich müde war, konnte ich doch nicht schlafen. Ich ging zur Themse hinunter und nahm ein Bad. Lange schwamm ich in dem lauen Wasser umher, mehr um nur etwas zu thun, als in der Erwartung mich davon besonders erfrischt oder gestärkt zu fühlen. Als ich wieder in's Haus trat, fand ich die Hausgenossen schon zu der ungewöhnlichen Stunde am Frühstückstisch versammelt. Aber es war in der That ein trauriges Zusammensein an jenem Morgen. Ausser der unsichtbaren Last, die uns alle drückte, hatte ein jeder noch seine eigne Sorge, die er nicht aussprach. Mein Vater fürchtete, dass seine Firma diesen Tag nicht überstehen würde. Meine Mutter war in grosser Sorge um meinen älteren Bruder, der mit seinem Regimente jetzt an der Küste stand; sie war zum Frühstück heruntergekommen, obwohl sie kaum im Stande war, ihr Zimmer zu verlassen. Meine Schwester Clara war am schlimmsten daran; sie hätte nicht einmal ihr ganz besondres Interesse für die Flotte zu verhehlen brauchen, denn wir alle wussten, dass ihr Herz dem jungen Lieutenant auf der Admiralsfregatte gehörte — einem der ersten Schiffe, die gesunken waren — aber es war nur eine stille Liebe gewesen und so wurde nicht darüber gesprochen; wir mochten dem armen Mädchen unsre Theilnahme nicht so unverhohlen aussprechen, wie wir sie empfanden. Dieses Frühstück — die letzte Mahlzeit, die wir noch zusammen einnahmen, war bald vorüber;

mein Vater und ich fuhren dann zur Stadt mit einem früheren Zuge als sonst, und kamen gerade dort an, als die trostlose Nachricht von dem Untergange unsrer Flotte von Portsmouth telegraphirt wurde.

Die Panic und die Aufregung an dem Tage war unbeschreiblich. Die *Consols* fielen von 93 bis auf 35; man bestürmte alle Banken; die Bank von England stellte sofort ihre Zahlungen ein; die Hälfte aller Häuser in der City erklärten sich fallit; die Regierung erliess ein Decret, welches alle grösseren Baarzahlungen suspendirte und zugleich allen Wechseln eine gesetzliche Verlängerungsfrist zusagte. Dies Moratorium aber kam zu spät für die meisten Firmen, und so auch für das Haus meines Vaters, *Graham & Co.* Sobald mein Vater in seinem Comptoir anlangte, erklärte er sich insolvent. — Wie dann ein *Aufruf zu den Waffen* erlassen wurde, und wie darauf das ganze Volk wie ein Mann aufstand — das Alles sind geschichtliche Thatsachen, die ich hier nicht zu erzählen brauche. Aber ich will euch wenigstens schildern, wie es mir persönlich dabei weiter ergangen ist.

Unser *Voluntier*-Wesen hatte ausserordentlich zugenommen seit dem Tage, dass der Krieg erklärt worden war; in ein paar Tagen vermehrte sich unser Regiment von seiner gewöhnlichen Zahl 600, bis auf über 1000; aber es fehlte an Gewehren. Man hatte uns eine Lieferung davon für den nächsten Tag versprochen, aber die haben wir nie mehr erhalten; und während wir auf diese Waffen warteten, wurde unser Regiment in zwei Abtheilungen getheilt: die

Rekruten exercirten mit unsern Gewehren des Morgens, und wir ältern Mitglieder des Regiments Abends. Die Fallissemente und die allgemeine Arbeitsstockung an jenem *schwarzen Freitage* hatten noch mehr Hände ausser Arbeit gesetzt, und fast jedermann liess sich nun bei den *Voluntier-Corps* anwerben. Am folgenden Tage war die Stärke unsres Regimentes schon bis auf 1400 Mann gestiegen, aber was konnten die machen ohne Waffen? An jenem Sonnabend freilich wurde angezeigt, dass eine Anzahl alter Musketen in der antiquarischen Waffenkammer des Tower an diejenigen Voluntier-Regimenter ausgeliefert werden sollten, welche Waffen verlangt hätten. In Folge dessen stürzten alle diese Voluntiers um die Wette nach dem Tower, und unsre Leute erlangten auch wirklich ein paar Hundert von diesen Gewehren; aber man hätte gerade so gut die Leute mit Besenstielen ihre Gewehr-Exercitien machen lassen können, als mit diesen ehrwürdigen Feuersteinschlossgewehren. Ueberdies war im ganzen Lande keine Munition für diese glatten Läufe mehr aufzutreiben. Es wurde zwar sofort eine National-Anleihe für die Fabrication von Hinterladern in Birmingham ausgeschrieben, und nahe an 50 Millionen Mark wurden in zwei Tagen gezeichnet, aber wie alles andre, so kam auch dies zu spät.

Unsre Truppen hatten schon seit 14 Tagen angefangen bei Dover, Brighton, Harwich und an andern Küstenplätzen Läger aufzuschlagen. In einigen derselben lag Miliz, in den meisten Linien-Truppen. Die Hauptquartiere unsrer Voluntier-Regimenter wurden jetzt auch in diese Läger verlegt, und die Voluntiers pflegten nun alle Tage per Eisenbahn zum Exerciren dorthin zu fahren; an jenem Freitage aber erging an die Voluntiers im ganzen Lande die Ordre, sich permanent zu den Waffen zu stellen. Nur die Voluntiers der Hauptstadt wurden als eine Art Reserve in der Umgegend von London gelassen, bis man sehen würde, an welcher Stelle die Invasion versucht werden würde; indessen waren auch wir schon in Brigaden und Divisionen einrangirt. Unsere Brigade bestand aus der *4ten Royal Surrey Miliz,* dem *1sten Surrey Verwaltungs-Bataillon* — wie man es nannte — in Clapham, dem *7ten Voluntier-Regiment* in Southwark, und meinem Regimente; aber nur das Bataillon und die Miliz lagen an demselben Orte einquartirt, und die ganze Brigade hatte nur zwei oder drei Nachmittage in Bushey-Park gemeinsam exercirt, ehe wir ausrückten. Unser Anführer sollte der Officier eines Linien-Regimentes in Irland sein, dieser aber stiess erst zu uns an demselben Morgen, als wir die Marsch-Ordre erhielten. Während der vorhergehenden 14 Tage hatte uns ein Oberst der Miliz commandirt. Aber obwohl wir Voluntiers es sehr wichtig hatten, mit unserm Exerciren und mit den Anstalten, die wir treffen wollten, so hatten doch alle die, welche, wie ich, in Bureaux der Regierung angestellt waren, mehr als genug dort zu thun, wie ihr euch wohl denken könnt. Diejenigen Angestellten, welche zu Voluntier-Corps gehörten, durften die Bureaux um 4 Uhr verlassen, die andern hatten bis tief in die Nacht hinein an ihren Pulten zu sitzen. Da waren die zahllosen Ordres für die verschiednen Regierungs-Aemter und die städtischen Behörden, Bekanntmachungen aller Art, Arrangements um die Werk- und Armenhäuser zu Hospitälern einzurichten; — diese und hundert andere Dinge waren in unserm Bureau auszufertigen, und da drinnen herrschte dieselbe Geschäftigkeit, wie draussen. Aber es war ein Glück für uns, dass wir so beschäftigt waren — die Leute, welche damals nichts zu thun hatten, waren jedenfalls am meisten zu bedauern.

Am Sonntag (das war der 15. August) arbeiteten wir ganz wie an den Werktagen vorher. Früh morgens ging ich zu einem Parade-Exerciren und fuhr dann mit dem 9 Uhr-Zuge in meiner Uniform zur Stadt; mein Gewehr nahm ich für alle Fälle mit und zum Glück auch meinen Regenmantel. Als ich im Waterloo-Bahnhof ankam, gingen dort allerhand unheimliche Gerüchte um. Von Dover aus wollte man eine Flotte gesehen haben und eines unsrer Aviso-Böte, die an der Küste hin- und herfuhren, brachte die Nachricht, dass eine Menge Schiffe unweit Harwich lägen; freilich waren sie von der Küste aus nicht zu sehen, aber vielleicht war die Luft nur zu unklar. Die Courier-Dampfer des Feindes hatten alle Fischerböte, die sie in Sicht bekamen, aufgefangen und versenkt, um zu verhindern, dass wir über ihre Stellung Nachricht erhielten: einige aber entkamen im Schutze der Nacht und erzählten unter anderm, dass unsre Fregatte Inconstant, welche von Nord-Amerika zurückgekommen war, ohne zu wissen, was sich inzwischen zugetragen hatte, in die feindliche Flotte hineingesegelt und dort sofort genommen worden sei.

In der Stadt machten sich alle Truppen marschfertig. Unsre Garde in der Wellington-Kaserne stand schon unter Waffen, und ihre Bagage-Wagen waren fertig gepackt. Die Parade-Schildwachen der Garde fand ich schon nicht mehr an ihrem Platze. Ordonanzen und Stabsofficiere sah man geschäftig nach allen Richtungen hin- und herlaufen.

An meinem Bureau arbeitete ich bis gegen 12 Uhr; und da ich seit dem frühen Morgen nichts gegessen hatte, so war ich in der That sehr hungrig und machte mich deshalb im Stillen fort nach meinem Club in Parliament Street, um etwas zu frühstücken. Es waren dort etwa ein halbes Dutzend Menschen im Speisesaal, die ich alle nicht kannte; wenige Augenblicke drauf aber kam Danvers, der am Finanz-Ministerium angestellt war, hereingestürzt, und von dem hörte ich endlich die ersten authentischen Nachrichten an jenem Tage. Der Feind hatte bereits bei Harwich eine starke Macht gelandet, und die Regimenter der Hauptstadt hatten Ordre erhalten, jene Truppen zu verstärken, welche dort in der Nähe stationirt waren; das Regiment, zu welchem Danvers gehörte, sollte um 1 Uhr antreten, und er war nur gekommen, um vorher noch etwas zu essen. Kaum aber hatten wir in aller Eile unser Frühstück verzehrt und wollten gerade vom Club fortgehn, als ein Bote vom Finanz-Ministerium gelaufen kam:

O, Herr Danvers, ich soll Sie holen, der Staatssecretär hat gesagt, dass alle Herren im Bureau bleiben sollen, und dass nicht ein einziger von Ihnen mit dem Regimente fortgehen darf.

Soll der Teufel holen, — sagte Danvers.

Wissen Sie, ob die Ordre sich auf alle Staats-Bureaux bezieht? — fragte ich den Boten.

Ich weiss es nicht, aber ich glaub's fast; ich hab' Boten nach allen Clubs und Restaurants laufen sehn, um nach den Herren zu suchen; der Secretär sagt, es sei ganz unmöglich, dass man einen einzigen von Ihnen jetzt entbehren könne, da ist soviel neue Arbeit zu thun gekommen; es sind Ordres eingelaufen, dass unsre Berichte heut Abend nach Birmingham geschickt werden sollen.

Ich nahm mir nicht die Zeit, Danvers zu beklagen, sondern sah nur Parliament-Street entlang, ob da auch keiner von unsern Boten zu sehen sei, der mich abfassen könne; dann lief ich so schnell ich konnte nach Westminster-Bridge zu und von da nach dem Waterloo-Bahnhof.

Da sah es schon ganz anders aus als am Morgen. Der fahrplanmässige Dienst hatte aufgehört; der Bahnhof und alle Zugänge waren von Truppen besetzt; hauptsächlich waren dies Garde und Artillerie. Alles ging in schönster Ordnung zu. Die Soldaten hatten ihre Gewehre zusammengestellt und standen nun in Gruppen umher. Man merkte dort nichts von Enthusiasmus oder Prahlerei; dazu waren die Zeiten doch schon zu ernst geworden. In jedem Gesichte sah man die Stimmung aller sich wiederspiegeln, und ein jeder fühlte jetzt, wie leichtsinnig wir die Warnung, die uns doch so eindringlich gegeben war, in den Wind geschlagen hatten. Nun war das Unheil, welches wir so lange als unmöglich und undenkbar verlacht hatten, doch über uns hereingebrochen und hatte uns unvorbereitet gefunden. Aber wenn jene Soldaten auch ernst dreinschauten, so waren sie doch ruhig und gefasst; man konnte sehn, dass sie entschlossen waren, ihre Pflicht zu thun, was auch immer kommen möge.

Ein Zug mit einigen Compagnien der Garde wurde gerade nach Guildford expedirt. Man sagte mir, er würde wohl in Surbiton anhalten, und so verschaffte ich mir mit noch mehreren andern Voluntiers von meinem Regimente einen Platz in diesem Zuge. Wir kamen dort auch nicht eine Minute zu früh an, denn unser Regiment kam in demselben Augenblicke schon auf der Landstrasse von Kingstone auf die Surbiton-Station zumarschirt. — Unsre Brigade hatte ihre Bestimmung nach der Ostküste erhalten, und sollte so schnell wie möglich von der Haupt-Güter-Station der Ost-Bahn in Shoreditch nach Harwich befördert werden, um dort sofort in Action zu treten. — Leere Wagen waren bald auf einem Seitengeleise zurechtgeschoben; unser Regiment sollte zuerst transportirt werden. Eine ungeheure Menschenmenge war dort versammelt um uns abfahren zu sehen, und unter andern auch die Rekruten, die bei uns während der letzten zwei bis drei Wochen gedient hatten, und die bei Weitem den grösseren Theil unsres Regimentes ausmachten. Sie sollten uns später nachkommen, wenn man sie nöthig haben sollte, aber sie waren uns eigentlich schon in jenem Augenblicke nur im Wege; denn da alle Officiere und Sergeanten zu unserm activen Theil des Regimentes gehörten, so war Niemand da, der unter diesen Rekruten hätte Ordnung halten können, und sie drängten sich um uns her, und zwischen unsre Reihen durch, und hinderten uns sehr beim Einsteigen.

Dort sah ich auch zum ersten Male unsern Brigade-General. Er war ein echter Soldat, der ohne Frage seinen Dienst kannte; aber für die Voluntiers war er nicht der rechte Mann; sie scheuten sich vor ihm, und er hatte nicht die rechte Art mit solchen Privatleuten umzugehen.

Ich wäre gern eben nach Hause gelaufen, um meinen Ueberzieher und den Tornister zu holen, den ich mir vor ein paar Tagen gekauft hatte, aber ich fürchtete zurückzubleiben. Ein gutmüthiger Camerad, einer von unsern Rekruten, bot mir an, die Sachen für mich zu holen, aber er war noch nicht wieder zurück, als unser Zug abfuhr, und so ging ich denn mit nichts als meinem Regenmantel und einem kleinen Gummibeutel voll Rauchtabak auf den Feldzug aus.

Der Zug war ganz ausserordentlich überfüllt; in jedem Coupé sassen zehn Mann, und es standen ausserdem noch drei oder vier; dabei war es ein drückend schwüler Nachmittag. Wir hatten unterwegs so oft anzuhalten, dass wir fast anderthalb Stunden brauchten, ehe wir im Waterloo-Bahnhofe anlangten; wir hätten beinah schneller zu Fuss hinkommen können. Es war zwischen 5 und 6 Uhr Nachmittags als wir dort ankamen, und es war fast 7 Uhr ehe wir vor der Shoreditch-Station aufmarschirten. Der ganze Bahnhof war überfüllt mit Proviant und Munition, die nach dem Osten geschickt werden sollten; uns wurde nichts davon gegeben. Wir stellten also unsere Gewehre mitten auf der Strasse zusammen und zerstreuten uns nach allen Seiten, um etwas zu essen und zu trinken aufzutreiben, was wir meist sehr nöthig hatten, namentlich das Trinken war einigen von uns, die von der Hitze und dem Gedränge elend geworden waren, sehr von Nöthen. —

Ich war eben im Begriff mit meinem Freunde Travers in eine Wirthschaft einzutreten, als gerade seine niedliche junge Frau in seiner Equipage angefahren kam. Die meisten von uns hatten schon in Surbiton von den Ihrigen Abschied genommen; sie aber wollte sich damit nicht zufrieden geben, hatte anspannen lassen und war mit ihrem prächtigen kleinen Jungen zur Stadt gefahren, um den Papa noch einmal zu sehen. Sie brachte ihm auch seinen Tornister und seinen Ueberzieher, und was noch besser war, einen grossen Korb mit gebratenen Hühnern, Zunge, Butterbrot, Cakes, und ein paar Flaschen Pontet Canet; — und diese wackren Leutchen bestanden darauf, dass ich mein Theil von diesen delicaten Leckerbissen abhaben müsste.

Die Zeit rückte vor, und der Abend senkte sich auf die Stadt. Die Surrey-Miliz, welche von Kingstone ganz zu Fuss durch die Stadt hermarschirt war, kam heran und inzwischen waren auch schon die andern Voluntier-Regimenter dort angelangt. Am Bahnhofe war es inzwischen etwas freier geworden; ein grosser Theil von den Vorräthen war bereits abgesandt; eine Abtheilung Artillerie, zwei Regimenter Miliz und ein Bataillon der Linie waren ebenfalls expedirt; nun sollten *wir* drankommen, und lange Reihen Waggons wurden schon für uns zurechtgeschoben; dennoch blieben wir in der Strasse aufgestellt. — Es schien in dem Augenblicke, als ob mehr Menschen in London seien, als je, und wir konnten vor lauter Zuschauern gar nicht vorwärts kommen — Höker, die Früchte verkauften, andre mit sonstigen nützlichen Dingen, Zeitungsjungen, alles drängte sich durch einander, von den Droschken und Omnibussen gar nicht zu reden; und dazwischen ritten Ordonanzen und Stabs-Officiere beständig ab und zu. Ein grosser Theil der Miliz-Soldaten und auch einige von unsern Leuten hatten offenbar zu viel getrunken; vielleicht war es auch die Wirkung der Sonnenhitze auf den leeren Magen; jedenfalls machten sie einen heillosen Lärm. Das Getöse, der Staub, und die Hitze den Augenblick waren ganz unbeschreiblich.

Immer weiter rückte der Abend vor; aber alle Informationen, die unsre Officiere aus unserm Brigade-General herausbringen konnten, waren die, dass wir für den Augenblick stille stehen und abwarten sollten. Unser General schien auf Instructionen von irgend einem andern General zu warten. — Nach und nach wurden die Strassen ruhiger und kühler. Der Brigade-General war, um uns ein gutes Beispiel zu geben, mehrere Stunden zu Pferde geblieben; man hatte ihm dann aus einem Laden einen Stuhl herausgeholt und nun sass der alte Herr auf demselben eingenickt. Die meisten unsrer Leute lagen oder sassen auf dem Strassenpflaster — einige schliefen, andre rauchten. Vergebens hatte Travers seine Frau gebeten, nach Hause zu fahren; sie blieb dabei, sie sei einmal so weit her gekommen, und wolle uns nun auch jedenfalls abfahren sehen. Die Equipage war in eine Nebenstrasse geschickt worden, da sie den Weg sperrte. Er hatte sich dann auf eine Thürschwelle gesetzt und sie sass neben ihm auf seinem Tornister. Der kleine Arthur war zuerst ganz ausser sich gewesen vor Vergnügen über den Lärm und die vielen Uniformen, aber als die erste Aufregung vorüber war, wurde er müde und eigensinnig, und weinte sich zuletzt in seines Vaters Armen in den Schlaf; sein gold-blondes Haar und sein kleiner dicker Arm hingen über dessen Schulter herab.

So war eine Stunde nach der andern traurig dahingeschlichen; dann plötzlich rüttelte uns ein Trompetensignal auf. — Wir sollten nach der Waterloo-Station zurückmarschiren. Die Landung an der Ostküste sei nur eine Kriegslist gewesen, hies es, der eigentliche Angriff würde vom Süden aus gemacht. — Irgend eine Entscheidung war uns lieber als die Ungewissheit und das Warten, und obwohl wir alle herzlich müde waren, so begrüssten wir doch die Order des Rück-Marsches mit grosser Freude. Wir liessen Frau Travers ihren Wagen selbst suchen. Der kleine Arthur, der nun wieder wach wurde, liess sich freundlich und ruhig von seiner Mutter forttragen.

Wir langten in Waterloo nicht vor Mitternacht an, und dann gab's wieder einen langen Verzug bis zur Abfahrt. Es waren dort einige Miliz- und Voluntier-Regimenter vom Norden angekommen; der Bahnhof und der ganze zur Station gehörige Bezirk war überfüllt mit Leuten, und es wurden Züge expedirt so schnell wie nur je in der Welt solcher Eisenbahndienst geleistet worden ist. — Diese ganze Zeit über hatten wir seit jener kurzen Anzeige keine weiteren Nachrichten gehört. Die Aufregung, welche zuerst geherrscht hatte, war allmählig unter dem Einflusse der Ermüdung und des Schlafbedürfnisses abgekühlt, und die meisten von uns schliefen schon, sobald wir nur im Zuge sassen. Ich selbst wenigstens schlief und wurde erst wieder aufgestört durch das plötzliche Anhalten unsres Zuges bei Leatherhead. Wir begegneten dort einem andern Zuge, der zur Stadt zurückkehrte, und es waren einige Leute in demselben, welche die neusten Nachrichten von der Küste mitbrachten. Wir konnten von unserm Ende des Zuges nicht hören, was sie sagten, aber das Gerücht lief schnell die Reihe der Wagen entlang: Der Feind war in Worthing gelandet und hatte seine ganze Macht bereits ausgeschifft. Seine Stellung war sofort von unsern Truppen, die bei Brighton lagen, angegriffen worden, und der Kampf sollte mit Tagesanbruch wieder beginnen. Die Voluntiers hatten sich gut gemacht. — Das war alles was wir erfahren konnten. Die Invasion war also doch geschehen! — Es war übrigens klar, dass der Feind nicht zurückgetrieben war, und so konnten wir noch zur rechten Zeit kommen, um ihn schlagen zu helfen.

Es war gerade im Augenblicke des Sonnenaufganges, als unser Zug langsam und vorsichtig in die South-Western Station bei Dorking hineinfuhr. Unzählige Male hatten wir schon unterwegs angehalten; hier aber blieb unser Zug eine lange Zeit stehen, und wir erhielten Erlaubniss auszusteigen, um uns etwas zu strecken und zu recken. Wir waren nur zu froh diese Gelegenheit zu benutzen, denn wir waren während der Nacht sehr eng zusammengestaut gewesen. Die meisten von uns machten hier ihr Frühstück von der Fourage, die sie sich von Shoreditch mitgebracht hatten. Ich hatte die Reste von

Frau Travers' Hühnern und etwas Brot in meinen Regenmantel gewickelt, und theilte dies jetzt mit ein paar weniger glücklichen Cameraden. — Wir konnten von der Station aus sehen, dass die Bahn vollständig mit Zügen besetzt war vorwärts so gut wie rückwärts.

Gegen 8 Uhr etwa erhielten wir Order wieder einzusteigen, und langsam bewegte sich endlich der Zug nach Horsham zu vorwärts. — Horsham-Junction, sagte man, sei der Punkt, wo wir Stellung nehmen sollten; aber als wir gegen 10 Uhr noch in einer kleinen Station ein paar Meilen vorzu hielten, kam der Befehl auszusteigen; und unsre Brigade formirte sich dann in Colonnen auf der Landstrasse. Hinter uns war eine Abtheilung Artillerie, uud vor uns (sagte uns ein Stabs-Officier) sei eine andere Brigade, die zusammen mit der unsrigen eine Division bildete.

Nach weiterem Verzug wurde unsere Colonne endlich in Bewegung gesetzt — aber nicht vorwärts: unsre Richtung war gen Nord-Westen, und eine bange Ahnung von dem wahren Sachverhalte dämmerte in meinem Geiste auf. Horsham war bereits von der Avant-Garde des Feindes besetzt; wir sollten uns auf die Anhöhen der Leith-Common zurückziehen und dort Stellung nehmen, um den Feind in der Flanke zu bedrohen, sobald er bis Guilford oder Dorking vorrücken sollte. — Diese Vermuthung bestätigte sich bald durch das, was der Brigade-General unserm Oberst sagte, und was wie ein Lauffeuer durch unsre Reihen ging. In jenem Augenblicke hörten wir zuerst Kanonendonner, der aus weiter Ferne von einer leichten Südbrise zu uns herüber getragen wurde.

Eine Stunde später hörte das Schiessen auf. Was hatte das zu bedeuten? — Niemand wusste es bei uns. — Mittlerweile setzten wir unsern Rückzug fort. Der Tag war wieder schwül und drückend; die Staubwolken, welche wir beim Marschiren aufwühlten, erstickten uns fast. Ich hatte noch eine Soda-Wasser-Flasche voll Rothwein von dem Abendessen in Shoreditch übrig; aber das reichte nicht weit, denn da waren viele durstende Kehlen, mit denen ich den Wein zu theilen hatte, und bald war der Durst schlimmer als vorher. Einige von unserm Regiment blieben thatsächlich vor Flauigkeit am Wege liegen, und wir mussten sehr häufig Halt machen, damit die Marodeure nachkommen konnten.

Endlich erreichten wir Leith-Hill. Dies ist der höchste Punkt in ganz Süd-England. Die Aussicht von dort ist wunderbar schön, und selbst an jenem Sommermorgen bot uns die Landschaft dort einen lieblichen Anblick dar, obwohl in der Nähe betrachtet, das Gras von der Sonne fast braun gedörrt war. Aber es war doch eine wahre Erholung für uns, von der staubigen Landstrasse endlich auf den Rasen der *Common* zu gelangen, und auf jenen Anhöhen erfrischte uns überdies eine köstliche Briese. — Dort konnten wir zum ersten Male unsre ganze Division übersehen. Unser eignes Regiment zählte nicht mehr als 500 Mann, da eine Menge Leute von den Regierungs-Bureaux dazugehört hatten, und die andren Regimenter waren meist nicht viel stärker. Das Miliz-Regiment aber war fast überzählig und unsre ganze Division, sagte man, seien im Ganzen nahezu 5000 Mann. Wir konnten von unserm Platze aus auch noch andre Truppen übersehen, die sich an unsre Division anschlossen; ich bemerkte ein paar Feld-Batterien unsrer Royal Artillery sowie auch einige schwere Geschütze, die offenbar einem Voluntier-Corps gehörten und von starken Arbeitspferden gezogen wurden.

Die kühlere Luft, das Bewusstsein unsrer starken Anzahl und die offenbar sehr gute Position, die wir eingenommen hatten, frischten unsern Muth wieder etwas auf; denn ich kann nicht leugnen, dass derselbe im Laufe des Morgen sehr gesunken war. Nicht, dass wir nicht voll Verlangen gewesen wären, mit dem Feinde handgemein zu werden, aber dieses Hin- und Hermarschiren und das ewige Haltmachen liess uns auf die Unentschlossenheit derjenigen schliessen, denen die Leitung des Ganzen anvertraut war. — Da war nun der Feind in zwei Tagen schon mehr als 20 Meilen weit in unserm Lande vorgedrungen, und noch war kaum etwas geschehen, ihn aufzuhalten. Ueberdies beunruhigte uns die Unkenntniss über das Vorrücken des Feindes, in der wir Voluntiers, von unserm Obersten abwärts, gehalten wurden. Wir malten uns die ganze Zeit über nur aus, wie der Feind seinen sorgfältig ausgearbeiteten Angriffsplan unbeirrt durchführte. Gerade diese Stille, mit der er vorzurücken schien, erfüllte uns mit einem geheimen Schauder.

Wie nun der Tag so in Unthätigkeit hinging, wurde uns sehr bald fühlbar, dass wir seit dem frühen Morgen nichts gegessen hatten. Aber keine Vorräthe irgend welcher Art waren zu haben, und Proviantmeister schienen überhaupt nicht da zu sein. — Als wir am Tage vorher vom Waterloo-Bahnhof abfuhren, stand dort ein ganzer Zug voll Vorräthen fertig aufgestellt, und unser Oberst schlug vor, man sollte einen Wagen dieses Zuges an den unsrigen anhängen, damit wir doch im Nothfall etwas zu leben hätten; aber der Officier, der an jenen Zug commandirt war — einen *Controlleur-Assistenten* glaube ich nannte man ihn — sagte, er habe Ordres, jene Vorräthe zusammenzuhalten, und ohne den Befehl seines Chefs könne er nichts von denselben fortgeben. So fuhren wir ohne den Proviant davon. Wer von uns Tabak hatte, rauchte seine Pfeife, und in der That giebt es keinen bessern Trost unter solchen Umständen. Ich hörte später, dass das Miliz-Regiment Proviant für zwei Tage in ihren Tornistern hatte, wir Voluntiers aber hatten keine Tornister und erhielten mithin auch nichts, was wir hätten hineinthun können.

Während dieser ganzen Zeit nun, dass wir dort im Grase lagen, und unsere Gewehre zusammengestellt hatten, sahen wir den commandirenden General und die beiden Brigade-Generäle mit ihrem Stabe langsam von einer Ecke der *Common* zur andern reiten, und beständig mit ihren Ferngläsern südwärts schauen; Ordonanzen und Stabsofficiere kamen und gingen ununterbrochen. Endlich ungefähr um 3 Uhr kam eine kleine Abtheilung Lanciers und ein Regiment berittene Miliz auf der Landstrasse von Horsham daher. Diese Reiterei schien weiter vorgerückt gewesen zu sein, und stellte sich jetzt in einer Colonne eine kurze Strecke vor unsrer Front auf, und zwar nach Süden gerichtet. Ob sie von da irgend etwas sehen konnten, weiss ich nicht; wir selbst waren am nördlichen Abhang der Anhöhe aufgestellt. Kurz darauf hörten wir ein Trompetersignal. Die Officiere wurden zum General befohlen, und erhielten kurze Instructionen; dann brach unsre Colonne wieder auf und marschirte wieder nach London zu weiter; dieses Mal bildete die Miliz den Nachtrab unsrer Brigade. Bald verbreitete sich ein Gerücht über diesen Rückmarsch durch unsere Reihen. Der Feind wolle uns dort nicht angreifen, sondern versuche unsre Stellung zu theilen; er formire eine Front gegen Reigate und eine andre gegen Aldershot, deshalb seien wir gezwungen zu retiriren und bei Dorking Stellung zu nehmen. — Die Höhen des grossen Kreide-Bergzuges also waren jetzt zu vertheidigen. Eine starke Macht ward um Guilford concentrirt, eine andere um Reigate; und wir selbst sollten Dorking auf uns nehmen, und dort den Feind erwarten. Das war, soweit wir Privatsoldaten in Erfahrung bringen konnten, die Operations-Basis für unsern Schlachtplan.

Wir marschirten also von Leith-Hill nach Norden zu abwärts. Von einigen Punkten hatten wir einen kurzen Blick auf die Bahn von Horsham nach Dorking, welche dort im Thale entlang läuft; wir sahen Arbeiter in rother Uniform auf derselben beschäftigt; es hiess, das sei das Genie-Corps, welches die Schienen aufnähme. Rastlos ging es voran. Der Staub schien schlimmer denn je. In einem Dorfe, durch das wir passirten — ich habe den Namen vergessen — stand eine Pumpe auf dem Dorfplatze. Hier endlich machten wir einen Augenblick Halt und tranken nach Herzenslust. Etwas später kamen wir durch ein grosses Bauerngehöft; die Pächtersfrau und zwei oder drei ihrer Mägde standen an der Pforte und reichten uns Stücke Brot und Käse aus Körben, die sie trugen. Ich erwischte auch einen Bissen von einem kleinen Stück; aber der Boden des Korbes dieser braven Frau war jedenfalls sehr bald erreicht. Von da an bekamen wir nichts mehr, bis wir endlich gegen 6 Uhr Abends in Dorking anlangten; die meisten Gutshöfe, durch die wir passirten, schienen schon verlassen zu sein.

In Dorking angekommen, stellten wir uns in Reihe und Glied in der breiten Hauptstrasse auf. Gerade uns gegenüber war ein Bäckerladen. Unsre Leute erbaten sich, anfangs zwei oder drei zur Zeit, die Erlaubniss, austreten zu dürfen, um sich dort Brot zu kaufen, aber bald liefen auch andre davon, und schaarten sich um den Laden, und zuletzt gab's eine richtige Balgerei dort. Wäre da irgend welche Ordnung gehalten, und eine regelrechte Vertheilung vorgenommen worden, so würden alle zweifellos Selbstbeherrschung genug gezeigt haben, aber der Hunger macht den Menschen egoistisch; jeder fühlte, dass sein Dahintenbleiben niemanden Vortheil bringen könne und dass vielmehr nur er sein Theil verlieren würde. So endete dieser Scandal in einer allgemeinen Prügelei des ganzen Regimentes. Der Laden war in zwei Minuten vollständig leer, und an's Bezahlen war bei dem Gedränge garnicht zu denken. Unser Oberst bemühte sich vergebens, dem Scandal zu steuern; einige der Officiere waren gerade so schlimm, wie die Gemeinen. In dem Augenblicke ritt ein Stabsofficier vorbei; er konnte kaum durch dies Gewühl hindurchkommen, und man drängte ihn in höchst unmanierlicher Weise hin und her. Er wurde mit Recht ärgerlich und rief uns zu: *So betragen Sie sich doch anständig, wie Soldaten, und nicht wie eine Bande Strassenräuber!*

O, warum nicht gar, Herr Prinzipal, rief Dick Wake ihm zu, Sie werden doch so ein paar armen Teufeln ihr bischen Futter gönnen? — Wake war erst kürzlich als Advocat immatriculirt, und war, wie wir es damals nannten, ein ganz impertinent gescheuter Junge, und übrigens gutmüthig genug. Als aber seiner schnodderigen Bemerkung noch einige faule Redensarten andrer folgten, wurde der Stabsofficier vollends aufgebracht.

Ordonanz, rief er dem Lancier, der hinter ihm ritt, zu, *führen Sie diesen Mann zum Profoss-Marschall. — Und was Sie betrifft, Herr Oberst, so würden Sie und Ihre ausgezeichneten Herren Officiere gut thun, dieses Pack etwas besser in Ordnung zu halten, wenn Sie nicht wünschen, dass ein paar von ihren Leuten noch vor ihrer Zeit erschossen werden.* — Unser Oberst sass dabei in stummem Erstaunen auf seinem Pferde, und der arme Dick, der schon begossen genug aussah, würde sicherlich am Schwanz des Sergeantenpferdes davon geschleppt worden sein, wenn nicht glücklicherweise noch unser Brigade-General dazu gekommen wäre, und die Erledigung der Sache auf sich genommen hätte. Er commandirte uns dann zur Stadt hinaus auf die Anhöhen an der Nordseite des Ortes. — Dieser Zwischenfall machte uns zugleich ärgerlich und auch verzagt. Es ärgerte uns, dass wir uns so wegwerfend hatten behandeln lassen müssen und doch fühlten wir, dass wir diese Behandlung verdient hatten; wir schämten uns unsrer Unmanierlichkeit. Dann aber hatten wir dadurch auch das Vertrauen zu unserm Oberst verloren, da er sich doch offenbar dabei blamirt hatte. Er war ein herzensguter Kerl, dieser Oberst, und bewies sich den nächsten Tag auch als ein braver Kerl; aber er suchte was darin, sich populär zu machen, und hatte eigentlich keine Ahnung davon, wie man Disciplin hält.

Wir waren übrigens kaum auf der Anhöhe angelangt, nach der wir hincommandirt waren, um dort für die Nacht zu bivouakiren, als wir hörten, dass ein Proviantzug am Bahnhofe angelangt sei. Fuhrwerk, um den Proviant zu uns herauf zu holen, hatten wir nicht; wir sandten also eine kleine Abtheilung aus, und diese Leute brachten uns die Nahrungsmittel in ihren Armen herauf: Bröte, ein Fass Rum, Packete Thee, und grosse Stücken Fleisch — genug für alle; dann aber hatten wir wieder keinen einzigen Kessel im Regiment, und wir konnten doch das Fleisch nicht roh essen. Der Oberst und die Officiere waren auch nicht besser dran. Sie hatten sich einen Regimentstisch arrangirt mit vollständigem Geschirr, Bedienung und allem, was dazu gehörte; aber dieses ganze Institut gelangte nie an Ort und Stelle, und was daraus geworden, oder wo es schliesslich hingerathen sein mochte, das ahnte niemand.

Einige von uns wurden nun in die Stadt geschickt, um zu sehen, was sie an Koch-Utensilien auftreiben könnten; ich war einer von diesen. Wir fanden die Strassen voll Artillerie, Train und berittnen Officieren, auch einige Voluntiers, die, wie wir, ausgegangen waren, um Einkäufe zu machen;

in allen Häusern schienen Soldaten einquartirt zu sein. Wir erlangten schliesslich wirklich ein paar Kessel und Töpfe; und ich verschaffte mir auch eine lederne Tasche mit einem Riemen dran, so dass ich sie um die Schulter hängen konnte. Dies Ding kam mir später sehr zu Statten. Mit diesen Schätzen beladen, schleppten wir uns zu unserm Bivouak auf der Anhöhe zurück, und füllten unterwegs noch die Kessel mit ziemlich schmutzigem Wasser aus einem Bache, den wir zu passiren hatten; denn bei demselben angekommen, fiel uns erst ein, dass oben auf der Anhöhe jedenfalls kein Wasser sei. — Die Entfernung war ein paar Meilen Weges, und wir waren schliesslich so erschöpft und überangestrengt, dass wir kaum im Stande waren, noch zu essen. Die Kocherei war überdies nicht gerade brillant, wie ihr euch wohl denken könnt. Alles was wir thun konnten, war Stücke Fleisch abzuschneiden, sie in den Töpfen zu kochen, und beim Essen unsere Finger als Gabeln zu benutzen. Wirklich erfrischend war allein der Thee, den aber tranken wir auch töpfenweise. — Es hatte übrigens unsern Leuten Mühe genug gemacht, überhaupt Feuer anzumachen, denn obwohl die Anhöhe mit wunderschönem Wald bedeckt war, so konnten wir doch kaum Holz genug zusammen bekommen, da wir nichts als unsre Taschenmesser hatten, um Zweige abzuschneiden.

Eben vordem es dunkel wurde kam der Brigade-General mit seinem Adjutanten zu unserm Oberst heran, um ihm zu zeigen, wo und wie er ein Piquet vor unsrer Position aufzustellen hätte. Ich glaube diese Vorsicht war nicht eigentlich nothwendig, da ja die Stadt vor uns lag und diese noch von unsern eignen Truppen besetzt war; aber zur Uebung mochte es wohl gut sein. Wir placirten dann eine Vorpostenlinie vor uns und hinter uns in Verbindung mit dem Regimente, das sich zur Seite an uns anschloss.

Endlich legten wir uns zum Schlafen nieder. Meine Compagnie hatte nicht die Wache, und so hatte ich wenigstens für die Nacht Ruhe; aber so müde ich auch war, so konnte ich bei der Aufregung in dieser ungewohnten Lage doch kaum schlafen; und obwohl die Nacht still und warm war und wir durch das Gehölz geschützt waren, fand ich es doch bald sehr kühl, da ich nichts hatte, um mich zuzudecken, als meinen dünnen Regenmantel, und um so mehr, da mein Zeug von Transpiration den ganzen Tag über nicht trocken geworden war. Nach einem sehr kurzen Schlummer wachte ich schon vor Tagesanbruch auf über und über fröstelnd und war froh, mich mit noch mehreren andern am Feuer wärmen zu können. Dann erst bemerkte ich, dass die Abhänge der Anhöhen an der Südseite der Stadt ebenfalls mit Wachtfeuern übersät waren. Wir glaubten anfangs es seien schon feindliche Feuer; bald darauf aber hörten wir, dass der Boden noch von einem starken Nachtrab unsrer Linien-Truppen besetzt sei, und dass wir keine Ueberrumpelung zu befürchten hätten.

Mit dem ersten Morgengrauen wurde Reveille geblasen; wir rappelten uns auf; dann wurde Appell gehalten. Ungefähr 20 Mann fehlten, die am Tage vorher krank liegen geblieben waren; ich glaube man hatte sie per Eisenbahn während der Nacht nach London geschafft. — Wir hatten schon etwa eine halbe Stunde in Reih und Glied gestanden, als der Brigade-General herankam und die Order gab, die Gewehre zusammen zu stellen. Wieder eine halbe Stunde später hiess es, wir sollten so schnell als möglich Frühstück machen und zugleich Essen für den ganzen Tag kochen. Dies bewerkstelligten wir dann ungefähr in derselben Art wie am Abend vorher.

Mittlerweile hatten wir Musse uns umzuschauen und die Lage zu erwägen, in der wir uns dort befanden. Unser Regiment stand an der äussersten Kante der Hügelkette, die sich von Guilford nach Dorking erstreckt. Diese ist zwar nur ein Theil des grossen Kreide-Bergzuges, der sich über eine viel weitere Strecke ausdehnt, von Aldershot im Westen bis an den Medway-Fluss im Osten; aber gerade da wo jener oben erwähnte Bach, der bei Dorking vorbeifliesst, die Mole, sich scharf nordwärts der Themse zuwendet, ist ein Thaleinschnitt in diesem Bergrücken, und an dem westlichen Abhange dieses Thaleinschnittes waren wir aufgestellt. Der Platz, wo wir bivouakirt hatten, schien ein herrschaftlicher Park zu sein. Etwas oberhalb dieser Stelle zu unsrer Rechten sah man ein schlossartiges Gebäude, zu dem der Park vermuthlich gehörte; dort hatte unsre Division ihr Hauptquartier aufgeschlagen. Von diesem Hause aus fiel der Abhang südwärts sehr steil ab nach dem Thal zu, das in nahezu ost-westlicher Richtung mit dem Bergrücken parallel läuft, und in welchem die Bahn von Reigate nach Guildford entlang führt. In diesem Thale, unmittelbar vor jenem Landhause in einer Entfernung von ungefähr anderthalb (engl.) Meilen lag die reizende kleine Stadt Dorking, wie in einem Nest von Bäumen versteckt, und lehnte sich an den gegenüberliegenden Thalabhang, der sich weiterhin bis nach Leith-Common, der Scene unsres Marsches am vorigen Tage, erstreckt. Die eigentliche Stadt Dorking lag also zu unsrer Rechten, aber die Vorstadt erstreckte sich so weit östlich, dass sie fast gerade vor uns lag; sie endete in einer der verschiedenen Eisenbahnstationen, die um Dorking herumlagen, und von dieser Station aus erhob sich der Park-Abhang, auf dem wir Stellung genommen hatten. Um diese Eisenbahn-Station gruppirten sich mehrere Landhäuser und ein oder zwei Wassermühlen, deren Gärten wir wie aus der Vogelperspective übersehen konnten; die kleinen Teiche glitzerten in der Morgensonne wie Spiegel.

Unmittelbar zu unsrer Linken in dem Thaleinschnitte, den ich eben beschrieben habe, längs der Mole und rechtwinkelig zu der Bahn von Guilford nach Reigate führte die Bahn von Horsham nach London über die Epsom-Downs, wo damals noch die weltberühmten Derby-Rennen alle Jahre stattfanden. Dicht bei dem Punkte, wo sich diese beiden Bahnen trafen, und auch nicht weit von dem eben genannten Bahnhofe war die Station der Waterloo-Bahn, auf der wir am Tage vorher unser Frühstück eingegenommen hatten.

An der andern Seite des Thaleinschnittes, also ostwärts (zu unsrer Linken) erhob sich die Fortsetzung des Kreide-Bergzuges. Die Schulter dieses Bergrückens nach dem Thaleinschnitte zu ist der bekannte Box-Hill, so genannt von den Buxbaum-Anlagen, mit denen die Anhöhe damals bedeckt war. Die Abhänge derselben sind sehr steil.

Der Kamm dieses ganzen Bergrückens war mit Truppen bedeckt; und die natürliche Stärke unsrer Position war auf den ersten Blick ersichtlich. Der steile Abhang, an dem wir aufgestellt waren, war meist nur mit Gras bewachsen und ohne Wald; das Ganze schien wie ein Schlachtfeld für uns gemacht zu sein. Unser schwacher Punkt war allein der Thaleinschnitt zwischen uns und dem Box-Hill. Konnte der Feind seinen Weg durch denselben forciren, so war unsre Schlachtlinie in zwei Theile geschnitten, und jeder dieser Theile konnte dann einzeln leicht überwältigt werden.

Aber ihr müsst nicht glauben, dass ich damals diese ganze Lage der Verhältnisse mit so kritischen Augen ansah. Jedem mussten unwillkürlich die evidenten Vortheile unsrer Position auffallen; was aber damals meinen Blick am meisten fesselte, das war die friedliche Anmuth der Gegend, die kleine Stadt, deren Häuser-Umrisse in blauen Morgendunst gehüllt waren, die üppige Fülle des Grüns und die massigen Formen der grossen schönen Bäume, der herrliche Sonnenschein, und der tief-blaue Himmel. Der Wald am südlichen Abhange des Thales von Dorking war so dicht, dass man ihn fast für einen Urwald hätte halten können. Der stille Frieden, der über der Gegend lag, zog mich um so mehr an, weil er so sehr mit der Vorstellung von dem contrastirte. was wir unmittelbar darauf erwarten mussten; und ich entsinne mich sehr wohl des bittren Unmuths, der mich überfiel bei dem Gedanken, dass es nun zu spät sei, diese kommende Verwüstung abzuwenden, und wie leicht wir sie doch hätten ganz vermeiden können. Jetzt war's zu spät! —

In jenem Augenblicke freilich schien Niemand diese gedrückte Stimmung mit mir zu theilen. Die Scenen in meiner unmittelbaren Nähe waren nichts weniger als düster; es war im Gegentheil Munterkeit und Lärm genug in unserm Lager. Wir hatten die erste Nachwirkung unsrer gestrigen Strapazen überwunden und hatten überdies wenigstens genug gefrühstückt. Wir waren voller Enthusiasmus, jetzt bald an der Vertheidigung unsres Vaterlandes activen Antheil nehmen zu können, und der Anblick der grossen Truppen-Macht, die wir um uns versammelt sahen, begeisterte uns. Am Abhang des Bergrücken entlang sahen wir noch fortwährend Truppen heranziehen — Voluntiers, Miliz, Cavallerie und Artillerie. Man sagte, diese seien am Tage vorher bis Leatherhead marschirt und jetzt seit Tagesanbruch von dort herübergekommen. Lange Eisenbahnzüge brachten auf der directen Bahn von London durch den Thaleinschnitt noch beständig Verstärkungen an Voluntiers und Miliz heran, welche meist auf dem Bergrücken zu unsrer Rechten, einige auch zu unsrer Linken jenseits Box-Hill aufgestellt wurden. Wir gehörten jetzt, wie man uns sagte, zu einem Armee-Corps, das aus drei Divisionen bestand; welche Regimenter aber die andern Divisionen bildeten, das habe ich nie erfahren.

Aufmerksam beobachteten wir alle diese Truppenbewegungen; wir hatten uns mit unserm Frühstück sehr geeilt, denn wir glaubten, die Schlacht könne jeden Augenblick beginnen. Am frühen Morgen hatten wir auch von Guilford einen Zug mit Rothröcken herüberkommen sehen; derselbe hielt an der kleinen Station zu unsern Füssen, und als die Truppen ausstiegen, konnten wir auch ihre Bären-Mützen erkennen. Es war ein Theil der Garde, der zu uns herübergeschickt wurde, um dieses Ende unserer Schlachtlinie zu verstärken. Sie liessen ein kleines Detachement am Eisenbahndamm zurück, der Kern der Abtheilung aber marschirte dann mit leichtem Schritt und zum Takte des lustigen Marsches ihres Musikcorps heran. Sie stellten sich quer über den Thaleinschnitt zu unsrer Linken auf, so dass sie dadurch unsre Schlachtlinie bis Box-Hill hinüber vollständig schlossen. Es schienen drei Bataillons zu sein.

Kurz darauf wurde ich nach Box-Hill hinübergeschickt mit einer Anfrage von unserm Oberst an den Oberst des Voluntier-Regimentes, das dort stationirt war. Es war bekannt, dass dieses Regiment besonders gut mit Wagen versorgt war, und wir hatten keinen einzigen; ich sollte nun um einen Ambulanzwagen bitten; meine Sendung war aber völlig nutzlos.

Als ich durch das Thal passirte, fand ich an der Eisenbahnstation eine ganz unbeschreibliche Verwirrung. Da kamen noch fortwährend Züge mit Proviant, Munition, Waffen und andern Dingen an, die alle ausgeladen wurden, so schnell wie es nur möglich war; aber es fehlten die nöthigen Transportmittel, um diese Vorräthe von dort weiter zu schaffen. Wagen waren allerdings da, aber keine Pferde um sie zu ziehen. Die Confusion dort wurde wesentlich vermehrt durch das Gewühl der Einwohner von

Dorking, die sich flüchten wollten; man hatte ihnen angezeigt, dass die Stadt wahrscheinlich ein Hauptschauplatz des Gefechtes werden würde. Damen und Frauen aller Arten und jedes Alters, und Kinder, von denen einige sich mit grossen Bündeln schleppten, jammerten nach Plätzen in dem Zuge, aber da schien Niemand authorisirt zu sein, diese Menschen fortzuschaffen, und die armen Geschöpfe drängten sich nun im Stationshofe hin und her, unaufhörlich aber ganz vergeblich um Auskunft jammernd und bettelnd, man solle sie doch fortlassen.

In dem Gedränge begegnete mir unser Regimentsarzt, der sich auch vergeblich bemühte, einen Ambulanz-Wagen zu bekommen; auch traf ich dort den alten Wood, Travers' Kutscher. Seine Herrin hatte ihn nach Guilford geschickt, weil sie glaubte, unser Regiment wäre dorthin beordert; er war zu Pferde gewesen und beladen mit einer Masse von Dingen, Proviant, wollnen Decken und natürlich auch einem Briefe. Er hatte meinen Tornister von Surbiton mitgenommen; aber in Guilford war sein Pferd für die Artillerie requirirt worden, und man hatte ihm nur eine Quittung über dasselbe in Händen gelassen. Diese Quittung aber konnte all die schönen Dinge nicht schleppen, so war er genöthigt gewesen, alles, was schwer war, zurückzulassen und unter andern auch meinen Tornister. Aber der alte treue Diener hatte doch mitgenommen, was er nur hatte tragen können, und als er hörte, dass wir hier in Dorking seien, war er zu Fuss von Guilford bis hierher gegangen. Er erzählte, der Ort sei überfüllt mit Truppen und die Anhöhen zwischen den beiden Städten seien vollständig mit Truppen besetzt; auch sagte er, dass einige Züge mit Verwundeten während der Nacht durch Guilford durchgekommen seien.

Ich nahm den Mann mit zu unserm Regiment und half ihm einen Theil seiner Sachen tragen. Den Proviaut hatten wir freilich jetzt weniger nöthig, aber die Teller und Messer, Gläser und dergl. schienen uns sehr wünschenswerth, und Travers, könnt ihr euch wohl denken, war besonders froh über den Brief. Ein paar Zeitungen, die der Alte mitbrachte, wurden von allen Umstehenden förmlich verschlungen, denn wir hatten seit Sonntag, als wir London verliessen, nichts Authentisches mehr gehört. Selbst jetzt nach so vielen Jahren, erinnere ich mich noch fast jedes Wortes, das ich damals las. Es waren zwei Exemplare von derselben Zeitung: das erste war am Sonntag Abend ausgegeben, als die Nachricht ankam, dass der Feind an drei Punkten zugleich mit Erfolg gelandet sei und war im Tone der äussersten Verzweiflung geschrieben. *Das Land sei überrumpelt. Der Sieger würde sich damit zufrieden geben, uns den Frieden in unserm eignen Lande zu dictiren. Es sei offenbar die Pflicht der Regierung, jetzt die besten Bedingungen anzunehmen, die man nur erlangen könne, ferneres Blutvergiessen und Unheil abzuwenden und das gänzliche Zusammenbrechen unseres wankenden Geschäfts-Credites zu verhüten.* — Die Ausgabe vom nächsten Morgen aber war schon in ganz anderm Tone geschrieben. Irgend etwas musste inzwischen zu unsern Gunsten geschehen sein, denn wir wurden darin wieder ermuthigt, Stand zu halten. *Unsre Position auf jenen Höhenzügen würde ganz uneinnehm-*

bar sein, und die Macht, welche wir dort concentrirten, sei der des Feindes an Zahl mehrfach überlegen. Die Invasion beweise sich schon jetzt als übereilt, und mit unsrer unbesiegbaren Schlachtlinie vor sich und der See hinter sich, bleibe dem Feinde keine Wahl, als gänzlich aufgerieben zu werden oder sich zu ergeben. Man solle doch nicht kleinmüthig von Unterhandlungen reden, der Kampf müsse ausgefochten werden; und er könne überhaupt nur einen Ausgang haben. England erwarte ruhig und mit voller Zuversicht den Angriff auf die unbesiegbaren Massen seiner Voluntiers. — Der Stil war sehr beredt, aber der Sinn war sehr schwach; vor allem lauteten auch die Nachrichten sehr widersprechend, denn ein paar Absätze weiter meldete dasselbe Blatt: *Die Regierung habe 500 Arbeiter von Woolwich nach Birmingham geschickt, um dort ein Zweig-Etablissement unsres Arsenals zu eröffnen.*

Eine ganze Zeit wurden wir damit in Bewegung gehalten, dass wir fortwährend unsre Stellung ändern mussten, einmal den Abhang höher hinauf, dann weiter zur Rechten, dann wieder tiefer unten zur Linken und so fort, je nachdem die Ordres an der Schlachtlinie entlang ausgegeben wurden; die Stabsofficiere galoppirten ab und zu, und das Gerassel der Artillerie, die von einem Ende des Bergrückens nach dem andern und wieder zurückgeschickt wurde, schien kein Ende nehmen zu wollen.

Endlich aber stand die ganze Schlachtlinie wirklich fertig aufgestellt, die Musik-Corps bliesen lustige Stücke und der General, der unser Armee-Corps commandirte, kam mit seinem Stabe herangeritten.

Wir hatten ihn schon ein paar Mal zu sehen bekommen, als wir vorher so oft unsre Stellung hatten wechseln müssen; jetzt aber hielt er eine Art formeller Inspection.

Er war eine lange schmächtige Figur und hatte sehr helles Haar; er sass sehr gut zu Pferde, und wenn man ihn so aufrecht im Sattel sitzen sah, wie er stolz unsere Reihen entlang ritt, hätte man ihn für 25 Jahre alt halten können; aber ich glaube, er war schon mehr als 50 Jahre activ, und war in Anerkennung seiner langjährigen Verdienste um die Armee zum Lord erhoben worden, als er schon ein alter Mann war. Er trug mehr Orden auf seiner Brust als da eigentlich Platz hatten; einige trug er wie eine Art Halsband umgehängt. Wie alle andern Generäle, so trug auch er eine blaue Uniform, mit einem Dreimaster-Hut und einem famosen Federbusch; das schien mir ein ganz alberner Firlefanz, denn es machte ihn sehr auffällig und konnte leicht den feindlichen Kugeln als Zielscheibe dienen.

Er hielt einen Augenblick vor unserm Bataillon an, und nachdem er eine Weile unsere Reihen entlang gesehen hatte, hielt er eine kurze Ansprache: *Wir hätten hier einen Ehrenposten nächst Ihrer Majestät Leibgarde und wir würden uns des Namens von Engländern würdig erweisen. Man brauche kein General zu sein, um die Stärke unsrer Position zu sehen; sie sei ganz uneinnehmbar, wenn sie nur einigermassen vertheidigt würde. Wir sollten warten, bis der Feind gut in die Enge getrieben sei, dann würden wir Befehl erhalten, draufzuschlagen. Ganz vor allem sollten wir nur fest Stand halten.* — Dann schüttelte er unserm Oberst die Hand, wir schrien ein Hurrah nach dem andern und damit ritt er weiter, zur Garde hinunter.

Nun also, dachten wir, wird die Schlacht losgehen. Aber vom Feinde war noch immer nichts zu sehen; und die Luft, obwohl sonnig und schwül, wurde doch so dick und unklar, dass man kaum die Stadt im Thale erkennen konnte. Von den gegenüberliegenden Hügeln waren nur unbestimmte Umrisse zu unterscheiden.

Nach einer Weile liess die Spannung, welche der Ansprache des Generals gefolgt war, nach; und wir schienen schon viel weniger das Gefühl zu haben, als ob alles davon abhinge, dass wir unsere Gewehre festhielten und sicher handhabten. Wir erhielten die Erlaubniss, die Gewehre zusammen zu stellen und in Schaaren von zehn bis zwanzig Mann in's Thal hinunter zu gehen, um in dem Bache zu trinken.

Dieser Bach und alle Hecken um denselben waren von unsern Plänklern besetzt. Die Stadt war wie ausgestorben. Dies konnte dem Feinde grossen Vortheil gewähren, denn wenn er sich erst einmal in der Stadt festgesetzt hatte, so war er dann besser gedeckt als wir.

Als ich unten am Bache war, kam plötzlich eine Colonne Soldaten aus der Stadt heraus, gerade auf uns zu. Wir dachten einen Augenblick, es sei der Feind; man konnte die Farbe der Uniformen vor lauter Staub nicht unterscheiden; aber bald stellte sich heraus, dass es unser Nachtrab war, der sich von den gegenüberliegen Anhöhen zurückzog, wo er die letzte Nacht campirt hatte. Ein Bataillon desselben hielt ein paar Minuten am Flusse an, um die Leute trinken zu lassen, und ich sprach einen Augenblick mit den Officieren.

Sie waren bei dem Corps gewesen, das den Feind bei seiner Landung angegriffen hatte, und sagten, sie hätten anfangs durchaus die Ueberhand gehabt, und würden den Feind leicht bewältigt haben, wenn sie nur irgend welche Unterstützung erhalten hätten, aber die ganze Sache sei von vorne herein verdorben. Die Voluntiers seien sehr dreist heranmarschirt, aber hätten sich sehr bald in Verwirrung aufgelöst; der Miliz sei es nicht viel besser ergangen; der Angriff sei vom Feinde mit Leichtigkeit zurückgeschlagen, und habe uns schwere Verlüste gekostet. Die Verwundeten dieser Truppen waren es gewesen, welche in der Nacht durch Guildford transportirt waren. Die Officiere fragten uns sehr angelegentlich nach der Zusammensetzung unsrer Schlachtlinie hier; als wir ihnen aber sagten, dass ein paar Bataillons Garde die einzigen Linientruppen auf dieser Flanke seien, schüttelten sie bedenklich den Kopf.

Während wir noch mit einander sprachen, trat ein dritter Officier zu uns heran; es war ein Mann mit schwarzem Haar, einem glatten Gesicht und eigenthümlich aufgeregten Manieren. *Sie sind Voluntiers meine Herren, nicht wahr,* sagte er sehr schnell und sperrte dabei seine Augen weit auf. *Glauben Sie nur, ich möchte Sie nicht gern verletzen, oder Ihnen auch nur irgend etwas Unangenehmes sagen; aber glauben Sie nur, wenn Sie alle sammt und sonders aufpackten und nach Hause gingen, so würde das ein verteufelt guter Streich sein. Ich versichre Sie, wir können diese Sache hier ganz allein ausfechten. Wir würden das sogar um ein Erkleckliches besser besorgen ohne Sie. Glauben Sie nur, wir brauchen Ihre Hülfe wirklich nicht. Wir möchten hier viel lieber allein sein, kann ich Ihnen sagen. Wissen Sie, ich möchte um keinen Preis unhöflich sein oder Ihnen zu nahe treten; aber es ist so, wie ich sage.* — Kaum war er mit dem Gepolter fertig, so war er auch schon wieder fort, ehe irgend einer antworten oder ehe die andern Officiere ihn aufhalten konnten. Sie entschuldigten sich übrigens wegen seiner Ungeschliffenheit, und erzählten uns, dass in ihrem Regimente auch ein Bruder von ihm gestanden habe, der am Sonntag gefallen sei; dieses Unglück, die Sonnenhitze und die Ueberanstrengung des Marsches hätten ihm den Kopf verdreht. Sie sagten uns auch, dass der Vortrab des Feindes ihnen schon dicht auf den Fersen gewesen sei, dass er aber noch zurückgeblieben sei, offenbar um auf Verstärkung zu warten, und dass er vor der Mitte des Tages wohl seinen Angriff nicht machen werde.

Es wurde sogar fast 3 Uhr Nachmittags, ehe die Schlacht begann. Unsre Geduld war mittlerweile bis auf's Aeusserste gespannt worden. Seit zwölf vollen Stunden hatten wir nun schon jeden Augenblick den Anfang des Kampfes erwartet, bis es uns zuletzt in einer Art von Ueberspannung fast vorkam, als ob diese ganze Invasion nur ein böser Traum sei, und als ob der bisher unsichtbare Feind wohl garnicht existire. Was wir bis dahin durchgemacht hatten, war nicht so sehr verschieden von unsern gewöhnlichen Voluntier-Manövers auf den Brighton-Downs, nur waren wir hier in grösserer Anzahl versammelt.

Ich entsinne mich noch wohl, dass mir damals solche Gedanken durch den Kopf gingen, als wir dort gruppenweise im Grase umher lagen; einige rauchten, andre kauten an ihrem Brotvorrath; manche schliefen sogar; und sorglos schien alles wie in Träumerei versunken. Da plötzlich wurden wir aus dieser Lethargie aufgerüttelt. — Ein Kanonenschuss war gefallen. Er kam vom Gipfel unsrer Anhöhe zur Rechten über uns und war nicht weit von dem erwähnten Landhause aus abgefeuert.

Es war der erste Schuss aus einer scharf geladenen Kanone, den ich je gehört hatte, und obwohl es 50 Jahre her ist, tönt mir noch das ängstliche Zischen dieses Schusses im Ohre. Dieses Geräusch sollten wir freilich bald nur gar zu sehr gewohnt werden. Alle sprangen unwillkürlich auf sobald sie den Knall hörten, und wir bedurften nicht erst des Befehles, um zu unsern Waffen zu laufen. Mit äusserster Anstrengung schauten wir nach dem Feinde aus.

Dieser erste Kanonenschuss war offenbar unser verabredetes Zeichen zum Beginn der Schlacht gewesen, denn unmittelbar darauf eröffneten alle unsre Batterien an den verschiednen Seiten unsrer Schlachtlinie ein heftiges Feuer. Wohin sie schossen, konnte ich nicht sehen, und ich bezweifle, dass die Kanoniere selbst viel mehr sehen konnten. Ich erwähnte schon, wie dick und unklar die Luft während des Morgens geworden war; jetzt legte sich gar der Pulverdampf wie eine Wolkenschicht über unsre Anhöhen. Wir konnten kaum unsre Reihen entlang sehen und konnten eben nur noch die Umrisse von ein paar Kanonieren einer Batterie unterscheiden, die unmittelbar neben uns am Abhange aufgestellt war.

Diese Kanonade dauerte, glaube ich, ein paar Stunden, und immer noch antwortete der Feind nicht. Wir sahen die Batterie in unsrer Nähe — es war eine Truppe reitender Artillerie — drauf los arbeiten, dass es eine Art hatte; der Officier, der das Commando führte, ritt langsam hinter seinen Leuten auf und ab, und stierte mit seinem Fernglas in den Dunst und Dampf hinein. Ein- oder zweimal hörten sie auf zu schiessen, um den Pulverdampf sich vertheilen zu lassen, aber das half nicht viel, da durchaus kein Wind war.

Ob eine Schlacht wohl so aussieht, wie dies hier, sagte Dik Wake, der mein Nebenmann war. *Mir kommt dies — wenig gesagt — zahm vor.* Aber er hatte das kaum ausgesprochen, so hörten wir schon ein Peloton-Feuer vor uns. Unsre Plänkler waren angegriffen worden, und sehr bald pfiffen auch schon die Kugeln über unsern Köpfen weg und einige schlugen dicht vor unsern Füssen ein.

Bis dahin hatten wir in Colonnen gestanden; nun aber wurden wir in Reihen commandirt und weitläufig über die ganze Strecke unseres Terrains vertheilt. Von dem Thaleinschnitte zu unsrer Linken führte ein Landweg zu uns herauf und vor uns entlang in fast genau westlicher Richtung. An der einen Seite dieses Weges war ein Redder aufgeworfen, ungefähr vier Fuss hoch, und hinter diesen hatte sich der grössere Theil unsres Regimentes placirt; aber etwas weiter den Abhang aufwärts bog der Weg ab, und dadurch wurde der rechte Flügel unsres Regimentes gezwungen, sich über den offnen Rasen des Parkes zu vertheilen. Meine Compagnie gehörte zu diesem rechten Flügel, der den Schutz des Redder-Walles entbehrte. Unmittelbar zu unsrer Rechten stand die Batterie, die ich schon erwähnte; dann kam ein Bataillon Linien-Truppen, dann wieder eine Batterie, dann eine Menge Miliz und Voluntiers und zuletzt noch wieder einiges Linien-Militair, welches bis an das erwähnte Landhaus reichte. So wenigstens war die Aufstellung, ehe das Schiessen anfing; was für Veränderungen seitdem vorgenommen sein mochten, weiss ich nicht.

Mit grossem Ungestüm eröffnete nun auch die feindliche Artillerie ihr Feuer. Wo sie ihre Batterien aufgefahren hatten, konnten wir nicht sehen, aber wir hörten deutlich das Sausen der Kugeln über unsern Köpfen und das Krachen, wenn sie hinter uns zersprangen. Was dann geschah, kann ich euch kaum recht beschreiben. — Manchmal wenn ich es versuche, mir diese Episode zu vergegenwärtigen, ist es mir als ob sie nur ein paar Minuten gedauert hätte; und doch kam es mir damals vor als ob die Stunden gar nicht enden wollten.

Wie im Banne eines bösen Traumes starrte ich beständig nach unsern Kanonieren hin, wie sie mechanisch ihre Arbeit fertsetzten. Unaufhörlich feuerten sie auf den unsichtbaren Feind; nur wenn ab und an mit einem dumpfen Schlage dort eine Kugel einschlug, trugen drei oder vier Cameraden die Verwundeten fort, aber wenige Augenblicke darauf waren sie schon wieder kaltblütig an ihrem Platze. Der Hauptmann ritt nicht mehr auf und ab; was aus ihm geworden ist, weiss ich nicht. Endlich hörten zwei von den Kanonen auf zu feuern, sie waren wahrscheinlich dienstuntüchtig geworden. Ein General der Artillerie kam herangeritten. Es ist als ob ich ihn noch heute vor mir sehe: ein sehr schöner Mann mit regelmässigen Zügen und schwarzem Schnurrbart; seine Brust war ganz mit Orden übersäet. Er schien in einer ungeheuren Aufregung darüber, dass die beiden Kanonen schwiegen;

Wer commandirt diese Batterie? — rief er.

Zu Befehl, Excellenz, antwortete ein Officier, der jetzt herzuritt, und den ich vorher nicht bemerkt hatte.

Wie ein Bild hob sich diese Scene scharf vom Hintergrund des Pulverdampfes ab: der General aufrecht auf seinem prachtvollen Renner, mit dem rechten Arm zeigt er auf den Feind, um dem jungen Officier etwas nachdrücklich zu erklären; dieser lenkt sein Pferd dicht zu ihm heran und salutirt mit der rechten Hand. Das war ein Moment, dann folgte ein dumpfer Schlag, und beide Reiter und Pferde lagen zu Boden geschmettert. Eine Vollkugel hatte gleichzeitig alle vier in der Sattelhöhe getroffen. Ein paar Kanoniere liefen herbei um zu helfen, aber keiner der beiden Officiere konnte kaum noch einige Minuten leben.

Dies waren nicht die einzigen, welche ich in meiner Nähe fallen sah. Fast unmittelbar darauf hörte ich etwas neben mir wie den Metallklang von Eisen, das auf Eisen schlägt, und im nächsten Augenblicke sah ich Dick Wake vorüberstürzen mit dem Gesicht auf die Erde; ein Schuss hatte ihm den rechten Schenkel abgerissen; die Kugel hatte dabei auf sein Bajonett geschlagen und das mochte wohl den Ton verursacht haben. Drei von uns trugen den armen Burschen fort mit grosser Mühe und Vorsicht wegen seines zerschmetterten Beines; aber er war schon halbtodt als wir bei dem Doctor anlangten. Dieser hatte sich in einer geschützten Vertiefung ungefähr 200 Meter weiter hinten placirt; auch waren dort noch zwei andere Doctoren in Civil, die freiwillig herzugekommen waren, um zu helfen. Der alte Wood machte sich dort sehr nützlich. Wir lieferten ihnen unsern Mann ab, und kehrten wieder zur Front zurück. Der arme Wake hatte noch volles Bewusstsein, als wir ihn verliessen, aber er war zu sehr erschüttert von dem Schlage, um sprechen zu können. Ich hatte noch mehrere solche Touren nach dem Orte zu machen, ehe dieser Tag zu Ende war.

Eine ganze Zeit lagen wir so und mussten auf uns schiessen lassen, ohne einen Schuss erwidern zu können; unsre Plänkler hielten noch ihre Position an den Dämmen und Büschen unten im Thal. Jetzt endlich commandirte unser Brigade-General unsern rechten Flügel auch hinter den Redder, hinter welchem die übrigen Leute unsres Regimentes postirt waren; wir lagen dort nun vier Reihen hinter einander. Ueber unsern Köpfen krachten die Bomben und pfiffen die Kugeln; aber kaum ein einziger von uns wurde dort verwundet. Unser Oberst allein war stark exponirt, denn er ritt auf dem Wege vor uns auf und ab so ruhig, wie wenn er von Stein gewesen wäre; den Major und seinen Adjutanten liess er absitzen und hinter dem Erdwalle Schutz suchen. Es that uns allen wohl, den Mann so gefasst zu sehen.

Die Zeit schien endlos, als wir so unthätig wartend dort versteckt lagen. Wir konnten es natürlich nicht lassen, ab und an über den Wall hinwegzugucken, um zu sehen, was unten vorging; aber man konnte nichts gewahr werden.

Es hatten sich inzwischen, ohne dass wir es bemerkten, schwere Gewitterwolken über uns zusammengezogen, und jetzt brach dieses Gewitter mit furchtbarer Gewalt los: ein jagender Wind, dann einzelne schwere Tropfen und endlich ein strömender Regen, der uns die Aussicht fast noch mehr verdeckte als vorher der Pulverdampf; das Aufflammen der Blitze und das Krachen des Donners mischten sich dominirend mit dem unausgesetzten Feuer der Artillerie.

Endlich hob sich der Regendunst einen Augenblick und wir konnten zu unsrer Linken sehen, wie schon der Feind am Box-Hill hinaufstürmte. Der steile glatte Abhang war übersäet mit den dunklen Uniformen. Die Front war eine unregelmässige Linie, aber die hinteren Reihen waren schnurgrade, compacte Massen. Der ganze Truppenkörper rückte in Absätzen vor; erst avancirte er eine Strecke, dann wurde gefeuert, dann weiter vorgerückt, dann gefeuert und sofort. Man sah die commandirenden Officiere ihre Säbel schwingen; eine Colonne nach der andern rückte heran und so setzten sie allmählig ihren Weg fort. Unsre Leute waren fast ganz versteckt hinter den Büschen auf der Höhe, von wo wir sie unaufhörlich feuern sahen. Plötzlich aber erschien vor diesen Büschen eine rothe Linie von unsern Gardetruppen und stürzte sich den Abhang hinunter auf den Feind. Dieser stutzte einen Augenblick, dann gab er nach;

offenbar wurde eine grosse Anzahl niedergemacht, und der Rest wich in ziemlicher Unordnung thalabwärts zurück. Dann zog sich der Wolkendunst nach jener Seite hin wieder dicker zusammen, so dass ich nicht mehr deutlich erkennen konnte, was dort vorging. Die Scene aber, die wir dort gesehen hatten, begeisterte uns alle, und wir hofften, dass wir dieselbe Bravour zeigen würden, wenn die Reihe an uns kommen sollte.

Mittlerweile waren unsre Plänkler zurückgeworfen worden und schienen bös zugerichtet; die meisten, welche ich sah, waren verwundet; einige hinkten allein, andre fanden noch Cameraden, die sie stützen konnten. Ihr Rückzug aber geschah doch in einiger Ordnung, und wir sahen einen Garde-Officier vor ihrer Front auf und ab reiten, der sie noch zu encouragiren suchte. Als sie vorbei waren, kamen endlich wir daran.

Einen Augenblick sahen wir vor uns nichts als grauen Wasserdunst, dann aber kamen uns knatternde Salven aus demselben entgegen; die Kugeln flogen jedoch meist über unsern Köpfen weg. Wir erwiderten das Feuer, aber dann kam unser Major zu uns herangeritten, und gab uns den Befehl, unsre Munition zu sparen. Wenige Minuten darauf wurden wir commandirt aufzustehen und dann konnten wir schon die Pickelhauben und endlich auch die Uniformen der feindlichen Linien erkennen. Der Feind schien sehr stark zu sein; sie maschirten wohl an fünf oder sechs Mann tief aber nicht mehr in geschlossenen Reihen, jeder von ihnen feuerte einzeln, wie wir es vorhin gesehen hatten, lief dann ein Stück vorwärts, feuerte wieder, und so rückte nach und nach die ganze Masse heran.

Unser Brigade-General kam jetzt den Landweg entlang geritten *Nun drauf, meine Herren, zeigen Sie was Sie können,* rief er uns zu, und wir schossen denn auch drauf los, was wir nur konnten. Ein wahrer Hagel von Kugeln aber zog auch über uns her und ich dachte, jeder Augenblick müsse mein letzter sein; aber in dem Momente sah ich Niemanden fallen, denn ich war zu sehr in Anspruch genommen, um rechts oder links zu sehen; ich schoss und lud und schoss nur ganz mechanisch weiter. Wie lange dies dauerte, weiss ich nicht, — sehr lange kann es kaum gewesen sein, denn unter solchem Feuer hätte keine der beiden Seiten viele Minuten ausdauern können. Das Ende aber war, dass der Feind langsam zurückwich, und sobald wir das bemerkten, erhoben wir ein ungeheures Geschrei; einige von uns sprangen auch auf den Erdwall und sandten ihnen noch einige Schüsse zum Abschied nach. Dann wurde plötzlich die Order ausgegeben das Feuern einzustellen; und wir wurden bald den Grund davon gewahr.

Ein Batallion der Garde marschirte von unsrer linken Seite aus quer vor unsrer Front vorbei, um dem Feinde in die Flanken zu fallen. Ich vermuthe, dass es mehr dieser Seitenangriff als unser Schiessen gewesen war, was den Feind zurückgeworfen hatte. Es war ein prachtvoller Anblick, diese geraden Linien unsrer Garde so stetig und regelrecht auf den Rasen vor uns abwärts marschiren zu sehen; sie schossen während sie vorrückten aber das Manöver wurde so exact ausgeführt, wie wenn es eine Parade gewesen wäre. Wir fühlten uns in diesem Momente so gehoben, als ob die Schlacht schon gewonnen wäre.

In dem Augenblicke rief Jemand aus, wir sollten uns der Verwundeten annehmen; da erst warf ich einen Blick unsre Reihen entlang, und sah die starken Verlüste, die wir erlitten hatten. Unmittelbar vor mir lag Bob Lawford, der auf demselben Büreau mit mir gearbeitet hatte. Eine Kugel hatte seine Stirn durchbohrt; im Todeskampfe hatte er noch krampfhaft sein Gewehr wie zum Schusse erfast. Ueberall sah ich Freunde und Bekannte tod oder verwundet. Ein paar Schritte abwärts fand ich Travers mit dem Rücken gegen den Erdwall sitzend. Eine Kugel war ihm durch die Lunge gegangen und das Blut strömte ihm aus dem Munde. Ich suchte ihn aufzuheben aber der fürchterliche Schrei, den er dabei ausstiess, liess mich ihn wieder hinsetzen. Ich sah dann, dass dies nicht seine einzige Wunde war; sein Schenkel war zerschmettert (die Kugel musste ihn getroffen haben als er auf den Erdwall gesprungen war) und das Blut aus seinen Wunden mischte sich mit den Pfützen von Regenwasser um ihn her. Aber er konnte unmöglich dort so liegen bleiben; ich hob ihn also auf, so vorsichtig wie ich konnte, und schleppte ihn langsam nach dem Feldhospital im Hintergrunde. Diese Bewegung muss ihm furchtbare Schmerzen gemacht haben, denn ich konnte den zerschossenen Schenkel nicht stützen, und er konnte sein Stöhnen nicht unterdrücken, ein so tapferer Soldat er auch immer war. Ich kann jetzt kaum begreifen, wie es möglich war, dass ich ihn überhaupt tragen konnte, denn er war viel grösser und schwerer als ich selbst. Ich war allerdings auch noch nicht sehr weit mit ihm gelangt, als mir schon der alte Wood und ein Mann vom Musikcorps mit einer Wegpforte als Bahre entgegen kamen; auf diese legten wir ihn. Wood hatte gerade noch Zeit mir zu sagen, dass er unten in dem Thaleinschnitte einen Wagen stehen habe, und versuchen wolle seinen Herrn sofort nach Kingston zu schaffen, als ich einen Stabs-Officier rufen hörte:

Sie müssen nicht so durcheinander laufen, meine Herren, bitte bleiben Sie doch in Reih und Glied.

Aber wir können doch unsere Verwundeten nicht so elend sterben und von den Hallunkon niedertreten lassen?! rief ihm einer zu.

Schlagen Sie erst den Feind mein Herr. — — Bitte, bitte, meine Herren, ich bitte sehr halten Sie sich zu Ihrem Regimente, wir sehen hier ja aus wie zusammengelaufenes Gesindel. — Und in der That sagte er das nicht ohne Grund, denn ausser unsern eignen Leuten waren auch noch eine Menge Voluntiers von andern Regimentern unseren Verwundeten zu Hülfe geeilt, so dass der ganze Platz mit einzelnen Gruppen von Menschen übersäet war. Indem ich dann nach meinem Platze lief, bemerkte ich noch wie hinter uns neue Massen von Truppen zusammengezogen worden waren, und eine Colonne marschirte zu unsrer linken in den Thaleinschnitt hinunter, wo vorher die Garde gestanden hatte.

Hatte nun während dieser Zeit das Gewehrfeuer nachgelassen, so war dagegen das Feuer der Artillerie nur stärker geworden. Die Bomben sausten über unsern Köpfen und platzten überall um uns her; ich kann nicht leugnen, dass ich froh war wenigstens den Schutz des Erdwalles wieder erreicht zu haben. Ueber denselben hinweg sah ich erst jetzt das furchtbare Unheil, welches unser Feuer vorher angerichtet hatte. Das Feld vor uns war fast ganz bedeckt mit Todten und Schwer-Verwundeten.

Eben über dieses Leichenfeld hinaus konnten wir noch in der Abenddämmerung die Bärenmützen und Rothröcke unserer braven Garde bemerken; sie waren unten am Abhange vertheilt und rückten dort noch langsam vor. Aber kaum eine Minute konnte vergangen sein, als unser Major zu Fuss den Landweg entlang kam (ich vermuthe dass sein Pferd gefallen war) und rief:

Zu den Waffen, meine Herren; sie rücken schon wieder an; und zum zweiten Mal sahen wir uns in einen Kampf verwickelt; es war wieder dasselbe Gewehrfeuer wie vorher. Wie lange das dauerte, weiss ich nicht zu sagen, aber ich erinnere mich noch deutlich der compacten feindlichen Linie ungefähr sechszig Schritt vor uns mit ihren Officieren zu Pferde. Wir schienen sie vollständig in Schach zu halten; sie waren unserm Feuer ungeschützt ausgesetzt, während wir fast bis an die Schultern gedeckt waren; aber ich fühlte dabei doch, das nicht Alles in Ordnung war. Plötzlich rief einer: *Wir werden von der Seite angegriffen!* — Ich blickte nach links; und sah im Dämmerlichte dunkle Gestalten über den Erdwall auf den Weg springen und unsere Front entlang feuern. Die Reserven des Voluntiers, welche die Stelle der Garde im

Thaleinschnitte eingenommen hatten, mussten vor dem Feinde gewichen sein; die feindlichen Plänker hatten uns offenbar von links umgangen.

Wie es dann kam, dass wir unsre Stellung veränderten, ob auf Commando oder vermöge natürlichen Instincts, weiss ich nicht; aber es dauerte nicht lange, so hatten wir alle den Landweg verlassen und formirten nothdürftig eine Front ungefähr 40 Schritte weiter zurück. Die Feinde besetzten jetzt den Erdwall von der andern Seite, und eine Anzahl von ihnen stürmte schon über denselben hinaus auf uns zu. Unser linker Flügel löste sich in Unordnung auf; die einzelnen Leute feuerten noch, indem sie sich zurückzogen, aber ganz unaufhaltsam rückte die geschlossne Front des Feindes dort vor. Einen Augenblick standen wir auf unserm Flügel noch und schossen auf's Gerathewohl in den Feind hinein, so schnell wir nur konnten. Unser Oberst und unser Major waren wohl gefallen; es war wenigstens Niemand da, der uns commandirte. Dann rief Jemand zu Pferde hinter uns (ich glaube, es war unser Brigade-General): *Nun drauf Voluntiers! Haut mit Hurrah auf den Feind ein! — Vorwärts!* — und mit wildem Geschrei stürzten wir uns auf die Feinde. Einige von ihnen liefen davon, die meisten aber hielten unserm Angriff stand, und nun gab's ein Handgemenge, Mann gegen Mann. Ich fühlte einen scharfen Stich im Bein gerade in dem Augenblicke, als ich einem Mann, der mir gegenüber stand, mein Bajonett durch den Leib stiess. Ich sah den armen Kerl rückwärts taumeln, seine Augen schienen ihm aus dem Kopfe herauszutreten, und waren wir auch den Augenblick wie zu wilden Thieren geworden, so ging mir dieser Anblick doch durch und durch. Aber der Kampf war im Nu vorüber; wir hatten den Platz bis an den Erdwall wiedergewonnen, und wäre nur jemand dagewesen, der uns hätte anführen können, wir wären auch weiter vorgerückt; aber wir waren bald wieder vollständig in Unordnung gerathen.

Die Feinde hatten den Erdwall sehr bald wieder besetzt, und Massen von ihnen stürmen an unsrer linken Seite vorbei. Wie es zuging, weiss ich nicht recht, aber wir waren nach und nach rechts die Anhöhe aufwärts zurückgewichen, von einer Front konnte garnicht mehr die Rede sein und unser linker Flügel, der schon früher nachgegeben hatte, flüchtete sich zu uns herüber, was die Unordnung bei uns nur noch vermehrte.

Es war jetzt fast vollständig dunkel geworden. — Auf der Höhe des Abhanges an der Seite, nach der hin wir uns zurückzogen, war eine starke Reserve in Colonnen aufgestellt. Einige in der vordersten Reihe derselben hielten uns für Feinde und feuerten auf uns. Unsre Leute riefen ihnen zu, einzuhalten und liefen auf sie zu. In wenigen Minuten aber war trotzdem der ganze Bergabhang eine Scene unbeschreiblicher Verwirrung; die Regimente und Detachements lösten sich in ganz hoffnungslose Unordnung auf. Die meisten von uns, glaube ich, wandten sich noch gegen den Feind, um ihre letzten Patronen zu verschiessen; es war jedoch zu dunkel, um zu zielen — glücklicher Weise für uns, sonst würden die Kanonen, welche der Feind durch den Thaleinschnitt heraufgebracht hatte, mehr Schaden unter uns angerichtet haben; so aber bemerkten wir kaum mehr davon als das Aufblitzen der Schüsse.

Durch unsre Verwirrung hatten wir einem Linien-Regimente unmittelbar hinter uns den Weg versperrt. Ich glaube, diese Truppen waren eben auf dem Schlachtfelde angelangt, und ihr Oberst und einige Stabsofficiere bemühten sich vergeblich, sich einen Weg durch uns hindurch zu bahnen; trotz des Kanonendonners und des wirren Durcheinanderschreiens, konnte ich doch ab und an ihr Rufen zu uns herüberschallen hören; sie commandirten uns, zur Rechten zurückzuweichen. Zuletzt gelang es einem berittnen Officier, sich durchzudrängen, und ihm folgte nach und nach eine Compagnie in Sectionen. Die Leute streiften an uns her mit ernsten und entschlossenen Gesichtern; sie waren sich der desparaten Aufgabe, die sie vorhatten, offenbar bewusst. Als das Bataillon sich durch uns hindurch gearbeitet hatte, schien es sich in Reihen zu formiren und den Abhang hinunter zu marschiren.

Lebhaft steht mir noch der letzte Angriff vor Augen, den zuletzt auch noch die Leibgarde machte. Sie ritt bei uns vorüber und warf sich auf die Stadt. Es war der letzte verzweifelte, aber vergebliche Versuch, die Schlacht zu retten, ehe wir das Feld räumten.

Endlich stiess der Adjutant unsres Regimentes, der in der Verwirrung von uns getrennt worden war, wieder zu uns, und führte uns, oder vielmehr diejenigen von uns, welche noch beisammen waren, auf die Spitze der Anhöhe.

Dort trafen wir eine wüste Menge von Voluntiers, Miliz und Train, alle in offenbarer Flucht begriffen. Wir wurden an dem Landhause vorbei von dem Menschenstrome wohl über eine Meile weit mit fortgerissen, ehe wir im Stande waren, anzuhalten. Zuletzt führte uns der Adjutant auf einen freien Platz etwas abseits von dem Zuge der Fliehenden; dort erst stellten wir wieder einige Ordnung unter den Resten unsrer Compagnien her. Dann hiess er uns warten und ritt davon, um Orders einzuholen, und den Rest unsrer Brigade aufzufinden.

Der Platz, auf dem wir uns befanden, war ein hochliegender Vorsprung der Berghöhe, und wir konnten von dem Punkte aus im Zwielicht der Sommernacht auf das Schlachtfeld hinuntersehen. Das Artillerie-Feuer war noch unausgesetzt im Gange. Wir konnten die Blitze der Kanonenschläge auf beiden Seiten sehen; ab und an kamen auch vereinzelte Bomben zu uns herauf geschwirrt und platzten nicht weit von uns; Gewehrfeuer aber konnten wir nicht mehr hören.

Dieser Halt hier liess uns zum ersten Male überdenken, was geschehen war. Obwohl wir damals noch nicht im ganzen Umfange begriffen, wie vollständig der Tag gegen uns entschieden war, so sagte uns doch unser Gefühl ungefähr, wie die Sachen lagen, und eine bittre Selbstanklage mag wohl in den Gemüthern der meisten von uns aufgestiegen sein; vor allem aber wurde uns jetzt erst klar, was für ein Verlust diese Schlacht für das Land war. Wir dachten an unsre armen verwundeten Cameraden, von denen wir keine Ahnung hatten, was aus ihnen geworden sein mochte, und zuletzt machte sich auch bei uns selbst die Reaction nach der Ueberanstrengung und Aufregung geltend. — Was mich selbst anbetraf, so bemerkte ich jetzt erst, dass ich ausser der Bajonettwunde am Beine auch noch eine Schusswunde am linken Arm hatte; eine Kügel hatte denselben dicht unterhalb der Schulter getroffen, aber ohne den Knochen zu verletzen. Ich entsann mich, etwas wie einen Schlag dort gefühlt zu haben, gerade als wir den Landweg verliessen; jetzt hatte die Wunde inzwischen schon zu bluten aufgehört und mein Hemd war an derselben festgeklebt.

Diese halbe Stunde kam mir wie ein Menschenalter vor, und während wir dort standen, erzählte uns der ununterbrochene Zug von Menschen und das Rasseln der Wagen, welche auf der Strasse den Bergrücken entlang flüchteten, auch ihre eigne Geschichte. Die ganze Armee war offenbar zurückgeworfen und der Feind durch den Thaleinschnitt vorgerückt.

Endlich konnten wir auch den Adjutanten unter der Menge erkennen, wie er aus dem Dunkel heraus zu uns her ritt. Die Armee solle sich zurückziehen, und auf den Epsom-Downs Stellung nehmen, sagte er; wir sollten uns dem allgemeinen Zuge anschliessen, und am andern Morgen versuchen, unsre Brigade wiederzufinden. So mischten wir uns denn wieder unter das Gewühl, und suchten so gut wir konnten, vorwärts zu kommen.

Der Adjudant theilte uns in einigen abgerissenen Sätzen den Stand der Dinge mit, als er neben unsrer vordersten Section herritt.

Wir hatten unsre Position eine Zeitlang gut gehalten, aber der Feind hatte zuletzt unsre Schlachtlinie zwischen uns und Guildford sowie auch in dem Thaleinschnitte neben uns durchbrochen, und hatte dann seine Leute durch die einmal gewonnenen Lücken vorwärts getrieben; dadurch war unsre Linie in Verwirrung gerathen, und auch das Armee-Corps in Guildford war dadurch genöthigt worden, sich zurückzuziehen, um nicht umgangen zu werden. Unsre Linien-Truppen bildeten gegenwärtig den Nachtrab, und wir sollten so schnell als möglich voraneilen, um ihnen Platz zu machen, damit sie am nächsten Morgen einen geordneten Rückzug antreten könnten. Der alte Lord, welcher unser Armee-Corps commandirt hatte, war schon am Anfange der Schlacht schwer verwundet vom Felde weggetragen worden. Die Garde hatte furchtbare Verluste gehabt. Unsre Garde-Cavallerie hatte die feindlichen Kürassiere niedergeritten, aber war selbst zuletzt entsetzlich zusammengehauen worden.

Das waren ungefähr die trostlosen Nachrichten, welche jetzt durch unsre Reihen liefen. Was aus unsern Verwundeten geworden war, wusste Niemand, und kaum mochte man danach fragen. — So schleppten wir uns müde und elend weiter.

Es muss wenigstens Mitternacht gewesen sein, als wir endlich L e a t h e r h e a d erreichten. Hier verliessen wir die Felder und hielten uns an die Landstrasse; damit aber wurde auch das Vorwärtskommen immer schwieriger. Die Passage war alle Augenblicke durch die grossen Menschenmassen versperrt; aber mühsam drängten wir uns doch voran. Einige Züge fuhren langsam vor uns her auf der Bahn, die dort der Landstrasse folgt; vermuthlich war es der Transport der Verwundeten oder doch wenigstens einiger derselben, deren man sich hatte annehmen können.

Der Tag war schon angebrochen als wir uns endlich Epsom näherten. Nach dem Gewitter war die Nacht klar und schön geworden, aber der kalte Nachtwind hatte mich in meinem vollständig durchnässten Zeuge durch und durchgeweht. Mein verwundetes Bein war ganz steif, und machte mir heftige Schmerzen. Ich war überdies vor Hunger und Erschöpfung im Begriff zusammenzubrechen, und meine Cameraden waren in nicht viel besserer Lage. Wir hatten seit dem Frühstück am Tage vorher nichts gegessen, und das Brot, welches wir uns aufbewahrt hatten, war von dem Gewitterregen fortgewaschen; ich fand nur noch etwas aufgeweichtes Mus in meiner Tasche. Mein Tabak war ebenfalls durchnässt und nicht zu rauchen. In diesem trostlosen Zustande schleppten wir uns langsam vorwärts, bis der Adjutant uns endlich abseits auf ein Feld führte. Erschöpft warfen wir uns dort in's Gras.

Hier wurde zunächst Appell gehalten, und nur 180 fanden sich noch vor von den 500, die wir am Morgen vor der Schlacht gewesen waren. Wie viele von den Fehlenden verwundet, wie viele getödtet waren, darüber konnten wir kaum Vermuthungen aufstellen; aber wir hofften, dass eine grosse Anzahl davon nur in der Verwirrung des Abends von uns getrennt worden sein würde.

Während wir noch da lagerten, sahen wir in dem Gedränge von Fuhrwerk und Menschen einen Wagen vorbeipassiren, der uns wie ein Proviant-Wagen vorkam, und von einem Mann in Uniform gefahren wurde. *Brot,* rief irgend Jemand, und instinctiv sprangen wohl ein Dutzend von uns auf und nahmen den Wagen in Beschlag. Der Kutscher versuchte uns mit der Peitsche abzuwehren; aber bald wurde er von seinem Sitz heruntergeworfen und der Inhalt des Wagens fortgeschleppt. Es waren Preserves in Blechdosen. Ohne weitere Umstände stachen wir dieselben mit unsern Bajonetten auf; ich vermuthe, dass das Fleisch vorher einmal gekocht worden war; jedenfalls schlangen wir es nur so hinunter.

Bald darauf kam ein General mit einigen Stabs-Officieren herangeritten. Er hielt still und sprach mit unserm Adjutanten; dann ritt er zu uns heran auf das Feld und sagte: *Ihr seid brave Jungen; ich will euch einstweilen in meine Division aufnehmen; schliesst euch dem Regimente, das dort passirt, an, sobald es vorbeimarschirt ist.* — Wir standen dann auf, und schlossen uns dem Zuge compagnieweise an, oder vielmehr in Abtheilungen von ca. 20 Mann stark; von da an gingen wir mit dem Strom, der sich auf der Landstrasse fortwälzte: — Regimenter, Detachements, einzelne Voluntiers oder Milizsoldaten, auch viele Landleute, die sich flüchteten, einige fuhren auf Leiterwagen, andre schleppten sich mit grosen Bündeln, die meisten liefen ohne alles zu Fuss daher; dazwischen sah man Wagen mit Vorräthen; und auf denselben sassen und hingen Leute, wo nur irgend einer Platz finden konnte; andre Wagen waren übervoll mit Verwundeten beladen. Ab und an entstand eine Stockung in dem Menschenzuge, wenn ein Pferd stürzte oder ein Wagen niederbrach und den Weg sperrte.

In Epsom selbst war die Verwirrung am schlimmsten; alle Häuser schienen voller Voluntiers und Milizsoldaten, die verwundet oder doch todtmüde und ausgehungert waren und die Strassen waren dicht gedrängt voller Menschen. Einige Officiere versuchten vergeblich Ordnung zu schaffen; es war ein hoffnungsloses Bemühen. Einige Voluntier-Regimenter, die während der letzten Nacht vom Norden angekommen waren, und hier Halt gemacht hatten, um weitere Befehle abzuwarten, standen längs der Landstrasse in schönster Ordnung, und einige der auf dem Rückzuge begriffnen Regimenter, so wie das unsre, mögen wohl den äusseren Anschein von Disciplin aufrecht erhalten haben, die grosse Menge Volks aber, die dort vorwärts drängte, war nur Pöbel. Unser Linien-Militär, oder doch alles, was noch von demselben kampffähig war, hatte sich, glaube ich, hinter uns gesammelt, um den Feind aufzuhalten.

Das allgemeine Gewühl wurde noch vermehrt, als man anfing, aus einigen Häusern die Verwundeten fortzuschaffen, damit sie nicht dem Feinde in die Hände fallen sollten; einige wurden auf Leiterwagen transportirt, andre wurden nach der Station getragen, um mit der Bahn befördert zu werden. Das Stöhnen und Wimmern dieser Unglücklichen anzuhören, wie sie durch die Strassen gerüttelt wurden, schnitt uns in's Herz, so gleichgültig uns eigne Noth und Erschöpfung auch sonst gemacht hatte.

Endlich wurde ein Entschluss gefasst und auf Anordnung eines Stabs-Officiers bogen wir von der Londoner Hauptlandstrasse ab und marschirten auf der Strasse nach Kingston weiter. Hier war das Gewühl etwas geringer, und wir konnten schneller vorwärts kommen. Die Luft war nach dem gestrigen Gewitter noch sehr kühl, und es war kein Staub. Wir passirten durch ein Dorf, wo unser neuer Brigade-General alle Wirthschaften hatte in Beschlag nehmen lassen, und sich in Besitz aller dort vorhandenen Spirituosen gesetzt hatte. Jedes Regiment musste, sobald es herankam, Halt machen, und jeder Soldat erhielt sein Glas Bier, das in Fässern compagnieweise vertheilt wurde. Ich hoffe, dass die Eigenthümer gut dafür bezahlt worden sind; aber es war für uns der wahre Göttertrank.

Es wird ungefähr 1 Uhr Nachmittags gewesen sein, als wir Kingston in Sicht bekamen. Wir waren nun sechszehn Stunden auf den Beinen gewesen und hatten zwölf Meilen zurückgelegt. — Etwas südlich von der Surbiton Station ist eine kleine Anhöhe, die damals zum grossen Theil mit Villen bedeckt war, ausgenommen am westlichen Ende, das von einem Haufen Bäume bedeckt war. Dahin hatten wir uns von der Landstrasse ab gewendet und hier liess unser General halten. Er stellte dann die ganze Division in einer Linie gegen Süd-Westen gewendet auf; der rechte Flügel reichte bis an die Wasserleitungs-Werke an der Themse und der linke Flügel dehnte sich den südlichen Abhang der Anhöhe hinunter bis an die Strasse von Epsom aus, auf der wir hergekommen waren.

Unser Regiment war ungefähr in der Mitte placirt, gerade dem General selbst gegenüber; dieser war abgestiegen, und hatte dort sein Pferd an einen Baum gebunden.

Die Anhöhe ist nicht sehr hoch, aber gewährt doch eine ziemlich weite Aussicht über die flache umliegende Gegend. Wie wir so dort im Grase lagen, konnten wir die Themse wie ein silbernes Band im hellen Sonnenscheine durch die Bäume glitzern sehen, an der andern Seite den alt-ehrwürdigen Palast von Hampton-Court, davor die Brücke bei Kingston. Der alte Kirchthurm des Städtchens erhob sich über den Dunstkreis, der auf den Häusern lagerte, und trat scharf hervor gegen den dunklen Ton des hinterliegenden Parks von Richmond.

In den meisten von uns rief dieser Augenblick wohl die Erinnerungen glücklicher Tage des Friedens, den wir genossen hatten, wach — Tage, die nun für immer dahin waren. Wir sprachen allerdings nicht mit einander davon; eine gänzliche Niedergeschlagenheit hatte sich aller bemächtigt. Diese war ohne Zweifel theilweise durch Ueberanstrengung und Mangel an Nahrung verursacht; ganz besonders aber auch dadurch, dass wir sahen, dass eine neue Schlacht versucht werden sollte. Für den Augenblick hatten wir alles Selbstvertrauen verloren. Hatten wir uns nicht in jener starken Position bei Dorking behaupten können, was für Chancen konnten wir dann wohl gegen denselben Feind hier auf diesem offnen Felde haben?

Dies Gefühl äusserster Verzweiflung brachte uns zu dem Entschluss selbst aller Hoffnung zuwider fortzukämpfen, aber zugleich kam uns jetzt doch der Gedanke an die Zukunft unseres Landes, an unsere Freunde und an alle die uns lieb und werth waren. Wir hatten keine Nachrichten über den Stand der Dinge im Lande, seitdem ich Wood am Tage vorher getroffen hatte. Wir wussten nicht, was in London inzwischen geschehen war, oder was die Regierung beschlossen hatte, und jemehr uns jetzt die Ueberlegung zurückkam, desto mehr erfüllte uns der Gedanke an andere und das brennende Verlangen zu wissen, was anderwärts vorging.

Unser General hatte erwartet hier Zufuhr von Proviant und Munition zu finden, aber bisher war nichts davon zu sehen. Die meisten von uns hatten kaum eine einzige Patrone nachbehalten; aber das Regiment, welches uns zunächst stand, und das bisher nicht im Treffen gewesen war, erhielt Befehl uns Patronen abzugeben, so dass wir wenigsten jeder 20 erhielten; und zugleich wurde eine kleine Abtheilung nach Kingston gesandt, um sich nach Proviant umzusehen. Einigen unserer Leute wurde auch gestattet auf eigene Hand in die Landhäuser hinter uns zu gehen und sich dort nach Nahrungsmitteln umzusehen. In ungefähr einer Stunde kamen sie mit etwas Brot und Fleisch zurück, so dass wir alle doch wenigstens etwas, wenn auch sehr wenig zu essen bekamen. Diese Leute erzählten auch, dass sie die meisten Häuser unbewohnt gefunden hätten, dass aber fast alle Vorräthe dort ausgeplündert und dass einige Häuser schon stark beschädigt seien.

Es muss wohl zwischen 3 und 4 Uhr gewesen sein, als wir in der Ferne vor uns wieder Kanonendonner hörten. Wir sahen den Pulverdampf über den Waldungen von Esher und Claremont aufsteigen.

Bald darauf kamen auch einige Truppenkörper auf der Ebene vor uns heran. Es war der Nachtrab unseres Linien-Militärs; auch wurden einige Kanonen herangefahren und ganz in unserer Nähe aufgestellt. Dies waren ursprünglich drei Batterien gewesen, jetzt aber waren davon im Ganzen nur noch acht Kanonen übrig. Die Brigade dieser Linien-Truppen bestand aus vier verschiedenen Regimentern, ihre ganze Anzahl aber war jetzt schwerlich mehr als acht- oder neunhundert Mann. Unser Regiment, und noch ein anderes hatte etwas zurücktreten müssen, um ihnen Platz zu machen. Etwas später wurden wir commandirt, die Eisenbahnstation (Surbiton) zu unserer Rechten zu besetzen.

Mein Bein war jetzt so steif geworden, dass ich nicht länger mit den übrigen marschiren konnte, und mein linker Arm war sehr geschwollen und schmerzhaft, so dass ich ihn durchaus nicht mehr anstrengen konnte; aber ich wollte alles andere lieber als zurückbleiben und so hinkte ich denn mühsam, so gut ich konnte, hinter dem Bataillon her bis zur Station hinunter.

Dort war ein Güterschuppen etwas weiter die Bahn hinunter, ein solides Mauersteingebäude; hier wurde eine Compagnie aufpostirt. Der Rest unserer Leute besetzte die Hofmauern des Bahnhofes.

Ein Stabs-Officier war mit unserem Regimente gegangen, um dort selbst die Anordnungen zu treffen;

wir würden eine Verstärkung von Linien-Militär erhalten, sagte er; und nicht lange darauf kam auch ein Zug mit solchen langsam von Guildford herangefahren. Das war der letzte Zug auf jener Bahn. Die Leute stiegen aus; der Zug fuhr in der Richtung nach London weiter, und dann wurde sofort eine Abtheilung Soldaten commandirt, die Schienen aufzureissen. Unter andern wurde uns in dem Schuppen auch ein Militärarzt beigegeben und ein Genie-Officier mit einigen Sappeurs, welche Löcher in die Mauer schlugen, die uns als Schiessscharten dienen sollten; freilich waren es nur ungefähr ein halbes Dutzend Leute, und so ging die Arbeit sehr langsam von Statten, da wir keine Werkzeuge hatten, um ihnen zu helfen.

Mittlerweile kam der Adjutant, der so frisch und thätig schien wie immer, zu uns in den Schuppen und commandirte uns im Stationshofe anzutreten. Die Abtheilung von Kingston war mit Proviant zurückgekehrt und wir hatten einen kleinen Bäckerkarren voll Brot, Mehl und rohem Fleisch als unsern Theil davon bekommen. Das Fleisch und das Mehl hatte ungekocht für uns keinen Werth, da wir kein Geschirr hatten, um sie zuzubereiten. Die Bröte aber verzehrten wir und fanden auch einen Wasserleitungs-Hahn im Hofe, so dass wir doch wenigstens mit dem Nöthigsten versorgt waren. Ich hätte auch sehr gerne meine Wunden gewaschen, die schon an zu eitern fingen, aber ich fürchtete meinen Rock auszuziehen, da ich ihn wahrscheinlich nicht würde haben wieder anziehen können.

Während wir all diese Vorbereitungen trafen, verbreitete sich bei uns das Gerücht von einem Unglück, das viel grösser war als irgend eines, das bisher geschehen war. Es hiess, Woolwich sei genommen. Wir alle wussten, dass dies unser einziges Arsenal sei, und begriffen die Tragweite dieses Schlages sehr wohl. Es war keine Hoffnung mehr, das Land zu retten, — wenn das wirklich wahr war. In äusserster Verzweiflung und Niedergeschlagenheit kehrten wir nach dem Schuppen zurück.

Obwohl dies erst unser zweiter Kriegstag war, so waren wir doch schon soweit alte Soldaten, dass wir gegen das Schiessen völlig abgestumpft waren, und die Kanonade von Kugeln und Bomben, die jetzt begann, machte keinen Eindruck auf uns. Wir fühlten in der That wohl unseren Mangel an Disciplin; aber wir waren dennoch alle entschlossen zu kämpfen, solange wir noch konnten. Die muthige Haltung unseres Adjutanten begeisterte uns; und der Stabs-Officier, welcher den Oberbefehl hatte, wusste sich auch das Air zu geben, als ob uns der schliessliche Sieg nicht fehlen könnte. Gerade als das Bombardement anfing, sah er zu uns in den Schuppen hinein und meinte, wir wären da so sicher wie in einer Kirche, und sollten den Feind nur tüchtig pfeffern, wenn er herankäme; wir würden sehr bald frische Zufuhr von Patronen erhalten.

Es fanden sich einige Tritte und Bänke in dem Schuppen; auf diese stellten sich unsere Leute, um durch die oberen Schiesslöcher zu feuern, und einige Soldaten von der Linie standen auf dem Boden parat, um die unteren Reihen zu nehmen. Ich setzte mich in der Mitte des Schuppens auf den Boden, denn ich konnte mein Gewehr nicht mehr halten und es waren ausserdem mehr Soldaten als Schiesslöcher da.

Das Artilleriefeuer, dessen Kanonenschläge wir hörten, kam von einem ziemlich entfernten Höhenzuge. Das Gewehrfeuer hatte noch nicht begonnen. Plötzlich hörten wir einen Krach im Schuppen, und ich wurde von einem Schlag am Kopfe zu Boden geworfen. Einen Augenblick war ich fast betäubt, und konnte zuerst nicht begreifen, was eigentlich geschehen war; dann aber wurde mir klar, dass wohl eine Kugel die Mauer des Schuppens getroffen haben musste, ohne durch dieselbe hindurchzudringen; aber der Schlag hatte die Bank, welche gegen die Mauer gelehnt war mit den Leuten, welche darauf standen, umgeworfen und hatte zugleich eine Wolke von Cementstaub und Stücke von der Mauer abgeworfen, von denen eines mich getroffen hatte. Ich sah ein, dass ich jetzt gänzlich nutzlos sein würde: ich konnte mein Gewehr nicht mehr halten und konnte kaum stehen; so beschloss ich endlich nach meinem Hause zu gehen, das nicht weit entfernt lag; vielleicht würde ich noch den einen oder andern meiner Hausgenossen dort antreffen, dachte ich. Ich stand auf und hinkte davon.

Das Gewehrfeuer war mittlerweile eröffnet worden und unsere Leute feuerten stark drauf los von den Fenstern der Häuser und hinter den Mauern, Wagen und wo sie sonst im Bahnhofe Schutz gefunden hatten. Ein paar Feldgeschütze waren in der Mitte des Hofes aufgefahren und auf dem freien Platze hinter der Station war ein Reserve-Corps aufgestellt. Dort war auch der Stabs-Officier zu Pferde, und beobachtete den Fortgang des Gefechtes durch sein

Fernglas. Ich hatte in dem Augenblicke noch genug Besinnung um einzusehen, dass unsere Position dort ganz hoffnungslos war. Es war klar, dass ein Feind wie der, mit dem wir es zu thun hatten, jene Reihe einzeln liegender Häuser sehr bald forcirt haben würde, und dann blieb unsern Truppen nichts übrig, als sich zu ergeben.

Es war ungefähr eine engl. Meile bis nach unserm Hause, und ich überlegte gerade, wie ich es möglich machen sollte, mich bis dorthin zu schleppen, als mir einfiel, dass ich das Haus, wo Travers wohnte, zu passiren hatte. Dasselbe lag in einer der ersten Reihen von Landhäusern an der Strasse, die von der Surbiton - Station nach Kingston führt. Ich dachte, ob er wohl zu Hause gebracht sei, wie sein treuer, alter Diener beabsichtigt hatte, und ob wohl seine junge Frau noch dort sei? Ich entsinne mich noch heute, wie ich mich schämte, dass ich an diesen, meinen besten Freund, seitdem er am Tage vorher blutend vom Schlachtfelde getragen war, kaum ein einziges Mal gedacht hatte. Aber der Krieg und die Noth machen den Menschen zum Egoisten. Ich entschloss mich zu ihm zu gehen und mich einen Augenblick auszuruhen. Vielleicht würde ich dort auch helfen können.

Der Garten vor dem Hause sah noch so hübsch und ordentlich aus wie immer. Ich war jeden Tag dort auf meinem Wege zum Bahnhof vorbeigegangen, und kannte jedes Blatt in dem Garten. Die Blumen blühten noch ebenso frisch wie sonst; aber es fiel mir schon auf, dass die Hausthür angelehnt stand.

Ich trat ein, und sah den kleinen Arthur auf dem Vorplatze stehen. Er war so niedlich angezogen, wie er es stets war, und wie er da so stand in seinem kleinen blauen Rock und mit seinem weissen Höschen und weissen Socken, welche die strammen, kleinen Beine sehen liessen, mit seinen gold-blonden Haar, seinem hübschen Gesichtchen und seinen dunkelblauen Augen, ein Bild kindlicher Anmuth; und dazu dies stille trauliche Haus, ganz so wie es sonst auszusehen pflegte, mit den Hüten und Röcken, die dort hingen, mit den bekannten Bildern und mit den Vasen voll frischer Blumen: — diese Oase des Friedens in Mitten des Krieges machte mich stutzen, und liess mich wie im wachen Traume zweifeln, ob doch der Teufelsspuk da draussen Wirklichkeit oder nur ein böser Traum sei. — Aber der Donner der Kanonen, der das Haus erbeben machte, und das Sausen der Kugeln gaben mir unzweifelhafte Antwort.

Der kleine Junge schien nichts von dem zu begreifen, was um ihn her vorging. Er kletterte gerade die Treppe hinauf, je eine Stufe zur Zeit, indem er sich am Geländer hinaufzog, wie ich ihn hundert Mal hatte thun sehn. Als ich eintrat drehte er sich nach mir um; meine Erscheinung aber erschreckte ihn; und wie ich so wankend zur Thür herein stolperte, das Gesicht und das Zeug mit Blut und Schmutz bedeckt, muss ich wohl für ein Kind ein grauenhafter Anblick gewesen sein; er fing an zu schreien und wollte die Kellertreppe hinunterlaufen. Als er aber meine Stimme erkannte und den freund-lichen Ton hörte, kam er schüchtern zögernd zu mir heran. Papa sei in der Schlacht gewesen, sagte er, und sei sehr krank; Mama sei oben bei Papa; Wood sei aus; Fanny sei unten im Keller und habe ihn auch dorthin gebracht, aber er wolle nun zur Mama.

Ich sagte ihm, er solle auf der Diele warten bis ich ihn riefe, stieg die Treppe hinauf und öffnete leise die Thür der Schlafstube. — Da lag mein armer Freund, sein Körper war auf dem Bette ausgestreckt, den Kopf an die Schulter seiner Frau gelehnt, die regungslos an seinem Bette sass. Er athmete schwer; die Blässe seines Gesichtes, die geschlossnen Augen, die krampfhaft starren Arme, der Schaum an seinem Munde: das alles erzählte vom herannahenden Tode, — Der treue, alte Diener hatte seinem Herrn den letzten Dienst geleistet; er hatte ihn doch wenigstens nach Hause gebracht, dass er in den Armen seines jungen Weibes sterben konnte. Das arme Wesen war zu sehr von ihrer Sorge und ihrem Schmerze in Anspruch genommen, als das sie mein leises Oeffnen der Thüre gemerkt hätte, und da das Kind den Augenblick besser von dort fern blieb, so schloss ich sacht die Thüre und ging wieder nach der Vordiele hinunter, um den kleinen Arthur im Keller in Sicherheit zu bringen, dort wo sich die Dienstboten des Hauses hingeflüchtet hatten. — Zu spät! — Ich fand ihn unten an der Treppe auf dem Gesicht liegen, seine kleinen Arme weit ausgestreckt und sein Haar im Blut schwimmend. Ich hatte während ich oben war den Krach nicht beachtet, aber ein Bombensplitter, der durch die offne Thür hereingekommen sein musste, hatte ihm den Hinterkopf vollständig abgerissen. Das Kind muss im Augenblicke todt gewesen sein. Ich versuchte den kleinen Leichnam mit meinem einen Arme aufzuheben, aber selbst diese Last war mir zu schwer; indem ich mich niederbückte verliessen mich die Kräfte, und ich verlor das Bewusstsein.

Als ich wieder zur Besinnung kam, war es ganz dunkel um mich her. Eine Zeitlang konnte ich nicht begreifen wo ich eigentlich war. Ich lag einige Zeit wie im Halb-Schlaf, und als ob ich mich nicht bewegen könne. Nach und nach wurde mir klar, dass ich auf den Teppich eines Zimmers lag. Alles Schlachtgetöse hatte aufgehört, aber es war ein Lärm wie von vielen Stimmen dicht bei meinem Ohre. Endlich richtete ich mich auf, und versuchte aufzustehn. Die Bewegung verursachte mir heftige Schmerzen, denn meine Wunden waren stark entzündet, und mein Zeug, das an dem Blute festgeklebt war, riss an denselben bei jeder kleinen Verschiebung. Zuletzt gelang es mir doch, ganz in die Höhe zu kommen; ich tastete um mich her und fasste an eine Thür, hart an der ich gelegen hatte; ich öffnete sie, und dann wusste ich auch schon wieder, wo ich war.

Ich hatte in Travers' kleinem Vorderzimmer gelegen und trat jetzt durch diese Thür auf die Diele hinaus. Das Gas war dort nicht angezündet; die Thür des gegenüberliegenden Empfangs-Zimmers war geschlossen, aber durch die halboffne Thür des nach hinten liegenden Ess-Zimmers fiel der schwache Schein eines Stearinlichtes auf die Diele.

In dem Ess-Saale sah ich wohl ein Dutzend fremder Gestalten. Es waren echte Niederdeutsche, — stämmige, kerngesunde Jungen mit gutmüthigen Gesichtern; aber dies Volk war damals noch nicht an solchen Wohlstand gewöhnt wie jetzt, und wie man ihn jener Zeit unter uns selbst beim niedren Volke fand. Der Tisch war mit Tellern, Gläsern und Flaschen bedeckt; die Mehrzahl der Leute schlief auf den Stühlen und auf dem Boden; einige rauchten und einer oder zwei, die noch ihre Pickelhauben aufhatten, waren mit Essen beschäftigt, und unterhielten sich dabei in abgerissenen Sätzen, die sie mit vollem Munde sprechend, so undeutlich hervorbrachten, dass es mir schwer wurde, sie zu verstehen, obwohl ich früher in ihrem Lande mit dem Dialect, den sie redeten, vertraut geworden war:

Na mit den Engelsmann is dat nu wol sacht to'n En'n, sagte einer von ihnen, ein breitschultriger Kerl, der gerade ein ungeheures Stück Ochsenbraten mit seinem Messer in den Mund stopfte. Zu welchem Zwecke er die silberne Gabel in der andern Hand hielt, war dem Manne offenbar noch nicht klar.

Jawol, ward ook Tied, erwiderte ihm ein Camerad, der sich auf die Long-Chaise hingestreckt hatte und überdies seine sehr schmutzigen Beine auf einen Stuhl stützte, der daneben stand. Er hatte eine von Travers' besten Cigarren im Munde. Sünd doch ganz fixe Kerls mang düt Volk, wenn se sick man nich so veel up eer Voluntiers inbill't har'n.

Na so slimm as de franszöschen Windhun'n sünd se doch nich.

Ne, aber 't is doch man so'n halben Kram; ick glöv nich, datt düsse Voluntiers hier gehörig inexerciert ward.

Dat am En'n nich; aber drapen könt se doch; — dat het uns hier veel Minschenleben kost.

Ja aber dat alleen deit et doch nich; dat Exercieren makt den Soldaten.

Wenn de Engelsmann doch man so'n paar Soldaten har, de wat lehrt hefft, denn har he nich mit uns Krieg anfangen müsst; wi har'n em ja nix dahn; aber he dacht sien Voluntier-Kram ward wol lang'n.

Weiter hörte ich nicht nach der schonungslosen Kritik hin, die diese Leute in ihrem einfachen Verstande an unsern verkehrten Verhältnissen übten, denn ein Geräusch auf der Treppe lenkte meine Aufmerksamkeit ab. — Frau Travers stand auf dem Treppenabsatze; ich schleppte mich zu ihr hinauf, so gut ich konnte.

Unter den vielen traurigen Bildern, die mir aus jenen Tagen in der Erinnerung haften geblieben sind, steht mir kaum eines so klar vor der Seele wie der schmerzliche Anblick dieser armen Freundin, die in wenigen Augenblicken ihres Gatten und ihres Kindes beraubt worden war; wie sie dort stand in ihrem weissen Sommerkleide, wie ein Geist, der aus dem Zimmer der Todten gekommen war. Sie hielt ein Licht in der Hand, dessen fahler Schein die Blässe ihres Gesichtes scharf hervortreten liess gegen das schwarze Haar, das ihr aufgelöst über die Schultern fiel. Ein edler Sinn leuchtete auf diesem Gesichte, wiewohl ihre schönen Züge von Mattigkeit und Kummer abgehärmt waren. Sie stand gefasst und ohne Thränen da; nur ihre zitternden Lippen liessen errathen, welche Anstrengung es ihr kostete, den namenlosen Schmerz zu bemeistern. — *Sie treuer Freund,* sagte sie, indem sie meine Hand fasste, *ich wollte eben kommen, um nach Ihnen zu sehen. Verzeihn Sie meinen Eigennutz, dass ich Sie so lange da habe liegen lassen; aber Sie begreifen wohl* (sie zeigte mit ihrem Kopfe nach oben) *dass ich in Anspruch genommen war.*

Wo ist . . . . fing ich an.

Mein Kind? — half sie mir aus mit dem, was ich zu fragen fürchtete. *Ich habe ihn zu seinem Vater gelegt. — — Aber jetzt lassen Sie mich nach Ihren Wunden sehen; o, wie blass und elend Sie aussehn! Setzen Sie sich doch nur hin!* — Damit ging sie in das Ess-Zimmer hinunter und brachte mir etwas Wein, den ich dankbar annahm; dann liess sie mich auf den obersten Stufen der Treppe niedersitzen, holte Wasser und etwas Leinewand, schnitt den Aermel meines Rockes ab, und wusch und verband meine Wunden. — Sie hatte sich vorhin eigennützig genannt; ich kam mir jetzt wirklich eigennützig vor, hier in dieser Weise ihr noch Sorge und Mühe zu machen; aber ich war thatsächlich zu schwach, um noch Willenskraft zu haben; ich bedurfte der Hülfe sehr, und sie zwang mich fast, dieselbe von ihr anzunehmen. Das Verbinden meiner Wunden gewährte mir auch in der That eine grosse Erleichterung.

Während sie sich so mit mir zu schaffen machte, erzählte sie mir in abgerissnen Sätzen, wie die Sachen standen. Alle Zimmer des Hauses ausser ihrem eignen, und dem kleinen Vorderzimmer, in das sie mich vorher mit Hülfe des alten Wood geschleppt hatte, seien von Soldaten in Beschlag genommen worden. Wood sei zum Dienst bei der Reparatur der Eisenbahn gepresst worden, und Fanny sei in ihrer Angst davon gelaufen; die Köchin sei glücklicher Weise an ihrem Platze geblieben, — habe den Soldaten das Essen gekocht und ihnen Travers' Weinkeller aufgeschlossen. — *Was die fremden Leute reden,* sagte sie, *kann ich nicht verstehen; sie sind recht ungeschliffen; — aber roh und gefühllos scheinen sie nicht zu sein; — um meine Sicherheit brauchen Sie sich nicht zu sorgen. Sobald ich Ihre Wunden verbunden habe, müssen Sie auch nach Ihren eignen Hausgenossen sehen — wie es dort nur gehn mag; man wird Sie dort gewiss sehr entbehrt haben; — ich — ich will hier Wache halten; — man wird mich dort wohl nicht belästigen.* Dabei warf sie einen kurzen und so schmerzlichen Blick auf die Thür des Zimmers, wo die Leichen ihres Mannes und ihres Kindes lagen.

Ich fühlte, dass sie Recht hatte. Ich war hier zu ihrer Beschützung überflüssig, und ich sehnte mich nach Hause, um zu erfahren, was aus meiner kranken Mutter und aus meiner unglücklichen Schwester geworden war; auch musste ich nothwendig Anstalten für die Todten in diesem Hause treffen. — Ich nahm kurzen Abschied, und schleppte mich dann fort. Es war nicht nöthig für mich, ihr viele Worte des Dankes zu sagen, und ihr Kummer war viel zu gross, als dass irgend ein Beweis meiner Theilnahme ihr hätte wohlthun können.

Auf der Strasse vor dem Hause war viel Lärmen und Treiben. Ein Zug von Wagen passirte die Strasse entlang; es war der Train der Sussex- und der Surrey-Regimenter, der offenbar dem Feinde in die Hände gefallen war und jetzt unter Bewachung der fremden Soldaten transportirt wurde. Auch auf der Strasse brannte kein Gas; aber die ganze Landstrasse nach Kingston zu war mit Fackeln erleuchtet, die von Menschen gehalten wurden. Diese waren in kurzen Entfernungen von einander aufgestellt, so dass sie eine lange Kette bildeten. Die Leute waren jedenfalls zu diesem Dienst gepresst worden, denn ich erkannte darunter einige Besitzer von benachbarten Landhäusern.

Einer der ersten von diesen Fackelträgern, der mir auffiel, war ein steifer alter Herr, dessen Physiognomie mir sehr bekannt war; ich war oft Morgens oder Abends mit ihm in demselben Zuge zur Stadt hin oder zurück gefahren. Ich glaube, er war Bureau-Chef einer Verwaltungs-Behörde. Er hatte ein zimperliches Wesen, pflegte ein sehr vornehmes Gesicht zu machen und trug seinen langen Hals für gewöhnlich in eine altmodisch-breite, weisse Cravatte gewickelt.

Trotz der Bitterkeit und des Ueberdrusses, die mich in dem Augenblicke erfüllten, musste ich doch über den Alten lachen. Es war eine zu absurd komische Figur, wie er da stand in seinem feinem Anzuge mit derselben feierlichen Miene und derselben ungeheuren Cravatte ganz wie sonst. Nun musste der alte Thor Busse thun für seine National-Unfehlbarkeit, — musste mit der schmutzigen Pechfackel in seiner Hand vor seinem eignen Hause den Eroberern die Strasse erleuchten. — — Aber ein wichtigerer Gegenstand nahm bald meine Aufmerksamkeit in Anspruch.

Ein Corporal mit ein paar Gemeinen führte auf dem Fahrwege zwei Voluntiers vorbei, denen die Hände auf den Rücken gebunden waren. Sie sahen mich mit flehenden Blicken an, und ich trat vom Trottoir hinunter, um den Corporal zu fragen, was die Leute gethan hätten. Dabei berührte ich seinen Aermel leise mit der Hand; sofort aber fuhr mich der Mann an:

Aus dem Wege hier, hob sein Gewehr, wie um mich fortzustossen und wandte sich zu den Gefangenen: *Vorwärts ihr Hallunken;* dann aber setzte er für mich, wie entschuldigend in gebrochnem Englisch hinzu: *Werden wir dies Volk von hinten auf uns lassen schiessen, sie jetzt geschossen werden;* und ich glaube, die beiden Unglücklichen würden wirklich erschossen worden sein, wenn nicht den Augenblick ein hochgestellter Officier in Begleitung mehrerer andrer zu Pferde herangekommen wäre. Diesem rief ich nun auf Deutsch an, so laut ich konnte:

Mein Herr, geschieht das mit Ihrer Bewilligung, dass diese wehrlosen Gefangenen erschossen werden sollen? —

Der Officier parirte sein Pferd und hielt den Corporal mit den Leuten an, um zu hören, was es gäbe. Meine Kenntniss der deutschen Sprache hatte mir wenigstens soweit geholfen, dass ich dadurch zu Gehör kam, während sonst wohl schwerlich auf mich geachtet worden wäre. Die beiden Gefangenen hatten sich natürlich nicht verständlich machen können, und hatten nicht einmal eine Ahnung davon, was sie eigentlich verbrochen haben sollten. Ich wollte gerade dem Officier ihre Erklärung übersetzen, aber kaum hatte ich angefangen zu sprechen, da unterbrach er mich. Er hatte sich inzwischen von dem Corporal den Fall kurz berichten lassen und wandte sich jetzt zu den Gefangenen, indem er ihnen in sehr gutem Englisch erklärte, worin ihr Unrecht bestünde. Diese versuchten dann ihm die näheren Umstände zu erklären: sie waren als Plänkler in der Nähe von Ditton in einer Scheune zurückgeblieben, und als sie endlich aus ihrem Versteck herauskamen, geriethen sie gerade unter einen Trupp feindlicher Soldaten. Da sie ihre Gewehre in der Hand hielten, hatte man geglaubt, sie hätten von hinten auf die Feinde schiessen wollen. Es war ein Wunder, dass sie nicht schon damals auf der Stelle niedergeschossen worden waren. — Der Officier liess die Leute in Freiheit setzen, und diese verschwanden hoch erfreut in einer Nebenstrasse.

Dieser Officier war eine noble Erscheinung und ein echter Soldat, aber seine Manieren waren ganz unbeschreiblich hochmüthig, und zwar um so mehr, da dieser Hochmuth unbewusst und unabsichtlich zu sein schien. Zwischen dem lahmen Voluntier, der ihn für seine gefangenen Cameraden anbettelte und ihm, dem stolzen Offizier des unbesiegten Heeres, war in seinen Augen eine ganz unendliche Kluft. Wären die zwei Gefangenen Hunde gewesen, so hätte ihr Schicksal kaum mit mehr Geringschätzung abgeurtheilt werden können. Was lag denn an dem Leben von so ein paar Vagabonden?! Aber es wiedersprach allerdings dem Gerechtigkeitsgefühl des Officiers überhaupt irgend welches Blut zu vergiessen, weder das eines Vagabonden, noch das eines Hundes, — wenn doch keine Ursache dazu gegeben war.

Doch was rede ich von solchem einzelnen Falle? — Jeder von uns hatte damals von Erniedrigungen und Demüthigungen, die ihm widerfahren waren, zu erzählen; da hörte man dieselbe Geschiche überall. Schon vom ersten Augenblicke an, als sie uns mit solcher Leichtigkeit in jener Schlacht über den Haufen warfen, und uns dann wie Gesindel vor sich hertrieben, war ihnen unsere ganze Macht nur lächerlich. Unsere Handvoll wirklicher Truppen war fast bis auf den letzten Mann aufgerieben worden; unsere Voluntiers und unsere Miliz aber, die ihren Dienst nicht kannten, und mit Officieren, die auch ihr Handwerk nicht verstanden, ohne Munition und ohne genügende Ausrüstung, ohne die nöthige Organisation und hungernd, wo doch Ueberfluss vorhanden war: die waren für eine Armee, wie die unseres Feindes nichts als hülfloser Pöbel — Freibeuter, die wohl hie und da desperat fochten, aber über die eine richtig organisirte und geschickt manövrirende Armee ohne Weiteres hinwegmarschirte, fast als ob wir garnicht dagewesen wären. — Glücklich waren die, deren Gebeine damals schon auf Surreys Gefilden bleichten; ihnen wenigstens blieb diese Schmach erspart. Selbst euch aber, die ihr jetzt es doch nie anders gekannt habt, als unter Entbehrungen zu leben, selbst euch brennt es in der Seele, wenn ich euch von jenen Tagen erzähle: was haben nun aber erst die erduldet, die wie ich in dem Bewusstsein aufgewachsen waren, der grössesten und stolzesten Nation auf Erden anzugehören, einer Nation, der niemals Schmach oder Demüthigung widerfahren war, die sich rühmte, eine Flagge zu führen, über der die Sonne nicht unterging, und deren Schlagwort war, dass sie den Ocean beherrsche, und dass sie nie anderen Völkern dienstbar werden könne!

Man redet wohl von der Generosität grosser Sieger — wir aber wurden nichts davon gewahr: Wir hätten den Krieg angefangen, hiess es, und wir müssten nun die Folgen tragen. London und unser einziges Arsenal waren in den Händen des Feindes und wir waren vollständig der Gnade des Eroberers Preis gegeben — seiner Gnade oder seinem Eigennutz; und konnte er wohl blind sein für die ungeheuren materiellen Vortheile, die seine Herrschaft über unser reiches Land in jenem Augenblicke ihm gewährten?! —

Was soll ich euch des Weiteren hier noch berichten! — von der ungeheuren Kriegsentschädigung, die wir zu zahlen hatten; von den Steuern, die uns auferlegt wurden, um diesen Betrag zu decken, und die uns bis heute noch in Armuth und in Elend niederhalten; — wie man uns damals grossartig erklärte, wir müssten jetzt eine neue Seemacht anerkennen, und wir müssten unschädlich gemacht werden; — wie die besten Schiffe unserer Flotte in die Marine dieser neuen Weltmacht übergingen; — wie die siegreichen Truppen hier im Lande auf unsere Kosten schwelgten, und wie das schwere Joch, das sie uns auferlegten dadurch so besonders unerträglich für uns wurde, dass es den Stempel von Methode und Legalität annahm. — Es war weniger schmerzlich seines Eigenthums von den Kriegsmanschaften mit Gewalt beraubt zu werden, als jetzt durch unsere eigenen Behörden, die zu Werkzeugen der Erpressung gemacht wurden. Wie wir solche Demüthigung, wie wir sie täglich, ja stündlich zu ertragen hatten, haben überstehen können, kann ich jetzt kaum begreifen. — Garnichts wurde uns von unserer Macht gelassen und damit ging uns auch die Basis unserer Existenz verloren. — Unserer Colonien wurden wir beraubt; Canada und die westindischen Besitzungen fielen an Amerika. Australien, Neu-Seeland und die Cap-Colonie wurden selbstständig. Irland riss sich von uns los und befindet sich seitdem in beständiger Anarchie und Revolution. Gibraltar, Malta, Cypern, Aden, Singapore, Hongkong und St.-Helena mussten wir der neuen Seemacht abtreten. Indien suchte sich lange noch ohne den Beistand dieses seines eigentlichen Mutterlandes selbstständig zu erhalten, und ist jetzt, wie ihr wisst, auch der neuen Weltmacht in die Hände gefallen.

Wenn ich jetzt sehen muss, wie unser Handel dahin ist, unsere Fabriken stille stehen und unsre Häfen leer sind, — das ganze Land nur eine Trutsätte des Pauperismus und ein enger Käfig, für den aufwärtsstrebenden Geist, der gewohnt war, im hohen Fluge über die Erde dahinzueilen — und wenn ich dann an meine Jugendzeit zurückdenke, wie solche Zustände damals vorzugsweise in dem Lande herrschten, das uns überwand, und wie damals Gross-Britannien jene erste Stellung in der Welt einnahm, die jetzt diesem andern Volke zugefallen ist: dann frage ich mich in der That, ob ich wohl ein Herz und Liebe für Alt-England habe, dass ich seinen Sturz so lange überleben konnte. Aber so vergeblich auch die Hoffnung war, dass die alten Zeiten wiederkehren möchten: diese Hoffnung war's allein, die mich aufrecht hielt. War doch Frankreich vordem gleiche Demüthigung von derselben Hand widerfahren, und es war dennoch nicht zu Grunde gegangen. Aber freilich sind und waren die Verhältnisse dort andre.

Frankreichs Sturz war kaum weniger unerwartet oder weniger verderblich als der unsre; den Franzosen aber konnte doch der reiche Boden ihres Landes nicht genommen werden. Die günstige Lage ihres weitausgedehnten Landes aber ist die eigentliche Quelle ihres zufriednen Wohlstandes. Colonien hatten sie kaum oder hingen doch in keiner Weise vom Besitze ihrer Colonien und von der wirthschaftlichen Ausbeutung fremder Welttheile ab: Sie konnten sich von solchem Schlage wohl erholen. — Anders war's dagegen seiner Zeit mit uns.

Unsre Prosperität hing allein von unserm Weltverkehre ab, von unserm auswärtigen Handel und von unserm weltbeherrschenden Credite. Sobald die Hauptwege des Handels einmal von uns abgelenkt waren, fehlten uns Macht und Mittel sie wieder an uns zu ziehen; und sobald unser National-Credit einmal erschüttert war, konnte keine Macht der Welt ihn wieder herstellen.

Wenn man in jener Alten Guten Zeit die Leute bei uns reden hörte, so schien gar Niemand zu bezweifeln, dass nach Gottes Rathschluss unser Governement für ewige Zeit müsse Anleihen zu 3 pro Cent machen können, und dass der Handel deshalb nur bei uns sich concentrirte, weil die kleine Nebel-Insel, welche wir bewohnten, rings vom Meer umspült sei. Man vergass, dass unser Reichthum von Amerika, Australien, Indien, Afrika und China herrührte, und dass die Leute, welche dort ihr Geld erworben hatten, aufhören würden sich im nebeligen England niederzulassen, sobald England aufhörte, der Mittelpunkt des Weltverkehrs zu sein. Niemand wollte glauben, dass die meisten unsrer Productionen anderwärts preiswürdiger würden hergestellt werden können als bei uns, und dass nur unsre künstliche Stellung im Mittelpunkt des Weltgetriebes uns productionsfähiger als unsre Nachbarvölker machte. Hatten auch wohl manche unsrer Vorzüge, wie der Reichthum unsres Landes an Kohlen und Eisen, unsre isolirte Lage und die Seemannschaft unsres Volkes uns besondre Vortheile gewährt, um unser Land zu solcher hervorragenden Stellung zu verhelfen, und zur ersten Macht der Welt zu erheben, so hätten wir uns doch sagen müssen, dass auch die grössten Vorräthe von Kohlen und Eisen nicht unerschöpflich, dass unser Land nicht unbezwinglich, und dass unsre Seemannschaft nicht einzig in der Welt sein könne. Aber wie damals unsre herrschende Weltstellung das Mittel geworden war, um den Reichthum der Welt bei uns anzuhäufen, so hätte diese Weltstellung allein auch dieses Mittel für uns bleiben können und sollen. Diese Stellung aber gründete sich lediglich auf unsre Weltmacht.

Dadurch, dass wir jene fernen Länder, welche unsern Reichthum producirten, stets mit unserm Lande in Verbindung hielten, dadurch, dass unser Volk sich eben jene Länder mehr und mehr zu eigen machte, und im wechselseitigen Austausch mit denselben auch daheim an Wohlstand zunahm, und dadurch, dass wir als die erste Macht der Welt jenen Ländern den sichersten Schutz und die grössten Vortheile gewährten: — Dadurch allein waren wir im Stande, die Welt commerciell zu beherrschen. Diese Macht aber versäumten wir uns zu bewahren und zu sichern.

Erst als es zu spät war, wurde uns klar: Wir hätten ebenso stark als Landmacht sein müssen, wie wir es als Seemacht waren, oder hätten andernfalls uns mit der stärksten Landmacht, der wir eben damals den Krieg erklärt hatten, zu Schutz und Trutz verbinden und mit ihr gemeinsam unsere Interessen in der Welt vertreten sollen.

J. F. Richter (H. O. Voigt's Buchdruckerei), Hamburg.

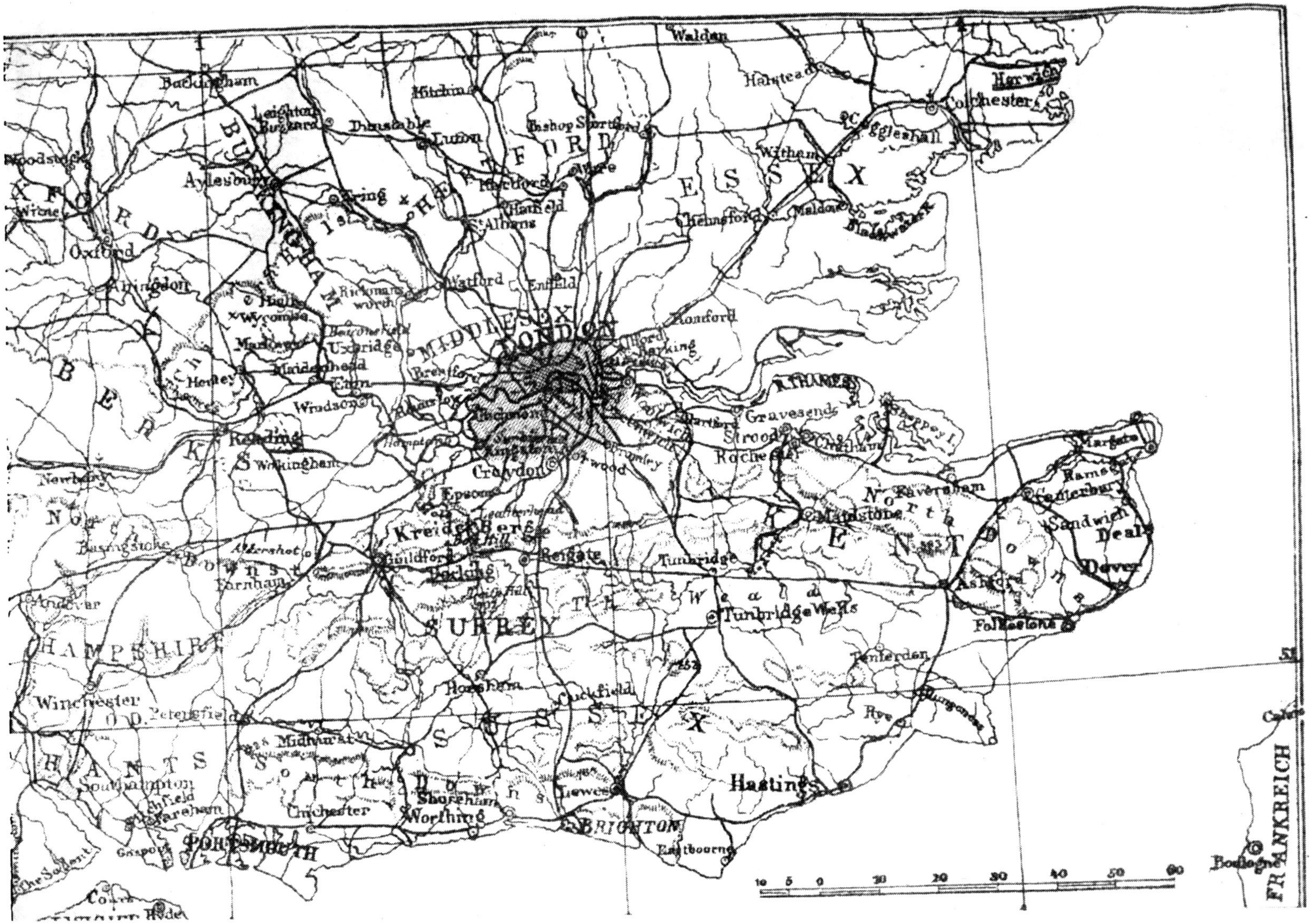
Walden
Buckingham
Hitchin
Halstead
Harwich
Colchester
Leighton Buzzard
Dunstable
Luton
Bishop Stortford
Coggleshall
Witham
Aylesbury
Oxford
Abingdon
Watford
Enfield
Romford
Barking
MIDDLESEX
LONDON
Uxbridge
Maidenhead
Eton
Windsor
Reading
Wokingham
Newbury
Basingstoke
Farnham
Guildford
Reigate
Epsom
Croydon
Kreide Berge
Gravesend
Strood
Rochester
Maidstone
Faversham
Canterbury
Margate
Sandwich
Deal
Dover
Folkestone
Ashford
Tunbridge
Tunbridge Wells
SURREY
HAMPSHIRE
Winchester
Petersfield
Andover
Southampton
Midhurst
Chichester
Worthing
Shoreham
Horsham
Cuckfield
Lewes
BRIGHTON
Eastbourne
Hastings
Rye
Tenterden
Hythe
Calais
Boulogne
FRANKREICH
10 5 0 10 20 30 40 50 60

LaVergne, TN USA
10 March 2011
219609LV00003B/17/P

9 781144 421456